AF449231

MOLECULAR BIOLOGY
INTELLIGENCE
UNIT

Nucleic Acid Switches and Sensors

Scott K. Silverman, Ph.D.
Department of Chemistry
University of Illinois at Urbana-Champaign
Urbana, Illinois, U.S.A.

LANDES BIOSCIENCE / EUREKAH.COM
GEORGETOWN, TEXAS
U.S.A.

SPRINGER SCIENCE+BUSINESS MEDIA
NEW YORK, NEW YORK
U.S.A.

Nucleic Acid Switches and Sensors

Molecular Biology Intelligence Unit

Landes Bioscience / Eurekah.com
Springer Science+Business Media, LLC

ISBN: 0-387-37491-4   Printed on acid-free paper.

Springer Science+Business Media, LLC, 233 Spring Street, New York, New York 10013, U.S.A.
http://www.springer.com

Please address all inquiries to the Publishers:
Landes Bioscience / Eurekah.com, 810 South Church Street, Georgetown, Texas 78626, U.S.A.
Phone: 512/ 863 7762; FAX: 512/ 863 0081
http://www.eurekah.com
http://www.landesbioscience.com

Printed in the United States of America.

9 8 7 6 5 4 3 2 1

Library of Congress Cataloging-in-Publication Data

Nucleic acid switches and sensors / [edited by] Scott K. Silverman.
    p. ; cm. -- (Molecular biology intelligence unit)
Includes bibliographical references and index.
ISBN-13: 978-0-387-37491-8 (alk. paper)
ISBN-10: 0-387-37491-4 (alk. paper)
1. Catalytic RNA. 2. Biosensors. 3. RNA--Biotechnology.
I. Silverman, Scott K. II. Series: Molecular biology intelligence unit (Unnumbered)
[DNLM: 1. Nucleic Acids. 2. Biosensing Techniques. 3. Genes,
Switch. 4. RNA, Catalytic--metabolism.   QU 58 N96438 2006]
QP623.5.C36N83 2006
572.8'8--dc22

                                        2006023015

About the Editor...

SCOTT K. SILVERMAN was born in Los Angeles, California and received his B.S. degree in chemistry from UCLA in 1991. He was an NSF and ACS Organic Chemistry predoctoral fellow with Prof. Dennis Dougherty at Caltech, graduating with a Ph.D. in chemistry in 1997. After postdoctoral research as a Helen Hay Whitney Foundation and American Cancer Society fellow with Prof. Thomas Cech at the University of Colorado at Boulder, he joined the University of Illinois at Urbana-Champaign in 2000, where he is currently Associate Professor of Chemistry. His laboratory focuses on nucleic acid structure, folding, and catalysis, using concepts and techniques from organic chemistry, chemical biology, and biochemistry.

CONTENTS

Section I: Artificial Nucleic Acid Switches and Sensors

EDITOR

Scott K. Silverman, Ph.D.
Department of Chemistry
University of Illinois at Urbana-Champaign
Urbana, Illinois, U.S.A.
Email: scott@scs.uiuc.edu

CONTRIBUTORS

Lieven P. Billen
Departments of Biochemistry
  and Chemistry
McMaster University
Hamilton, Ontario, Canada
Chapter 4

Ronald R. Breaker
Department of Molecular, Cellular
  and Developmental Biology
Yale University
New Haven, Connecticut, U.S.A.
Email: ronald.breaker@yale.edu
Chapter 6

Boris François
Institut de Biologie Moléculaire
  et Cellulaire du CNRS
Modélisation et Simulations des Acides
  Nucléiques, UPR 9002
Université Louis Pasteur
Strasbourg, France
Email: B.Francois@ibmc.u-strasbg.fr
Chapter 7

Yoshiya Ikawa
Department of Chemistry
  and Biochemistry
Graduate School of Engineering
Kyushu University
Fukuoka, Japan
Chapter 3

Tan Inoue
Graduate School of Biostudies
Kyoto University
Kyoto, Japan
Email: tan@kuchem.kyoto-u.ac.jp
Chapter 3

Edward K.Y. Leung
Department of Molecular Biology
  and Biochemistry
Simon Fraser University
Burnaby, British Columbia, Canada
Chapter 2

Yingfu Li
Departments of Biochemistry
  and Chemistry
McMaster University
Hamilton, Ontario, Canada
Email: liying@mcmaster.ca
Chapter 4

Oliver Mayer
Department of Microbiology
  and Genetics
University Departments
  at the Vienna Biocenter
Max F. Perutz Laboratories
Vienna, Austria
Chapter 5

Ali Nahvi
Department of Molecular Biophysics
  and Biochemistry
Yale University
New Haven, Connecticut, U.S.A.
Chapter 6

Razvan Nutiu
Departments of Biochemistry
    and Chemistry
McMaster University
Hamilton, Ontario, Canada
Chapter 4

Renée Schroeder
Department of Microbiology
    and Genetics
University Departments
    at the Vienna Biocenter
Max F. Perutz Laboratories
Vienna, Austria
Email: renee.schroeder@univie.ac.at
Chapter 5

Dipankar Sen
Department of Molecular Biology
    and Biochemistry
Simon Fraser University
Burnaby, British Columbia, Canada
Email: sen@sfu.ca
Chapter 2

Garrett A. Soukup
Department of Biomedical Sciences
Creighton University School of Medicine
Omaha, Nebraska, U.S.A.
Email: gasoukup@creighton.edu
Chapter 1

Quentin Vicens
Department of Chemistry
    and Biochemistry
Howard Hughes Medical Institute
University of Colorado
Boulder, Colorado, U.S.A.
Email: vicens@colorado.edu
Chapter 7

Herbert Wank
Department of Microbiology
    and Genetics
University Departments
    at the Vienna Biocenter
Max F. Perutz Laboratories
Vienna, Austria
Chapter 5

Eric Westhof
Institut de Biologie Moléculaire
    et Cellulaire du CNRS
Modélisation et Simulations des Acides
    Nucléiques, UPR 9002
Université Louis Pasteur
Strasbourg, France
Email: E.Westhof@ibmc.u-strasbg.fr
Chapter 7

Nikolai Windbichler
Department of Microbiology
    and Genetics
University Departments
    at the Vienna Biocenter
Max F. Perutz Laboratories
Vienna, Austria
Chapter 5

Switches and sensors composed of nucleic acids are being developed in the laboratory and have also been identified in nature. In this book, seven chapters describe studies aimed at understanding and exploiting the key features of such molecular RNA and DNA devices. In the first section of the book, four chapters are devoted to **artificial nucleic acid switches and sensors**. These chapters introduce the concept of allosteric ribozymes as molecular switches and sensors; describe nucleic acid enzymes that are switched by oligonucleotides and other nucleic acid enzymes that are switched by proteins; and illustrate how switching elements can be integrated rationally into fluorescently signaling molecular sensors made out of nucleic acids. In the second section of the book, three chapters show that nature has been as crafty a molecular-scale engineer as any modern scientist via evolution of **natural nucleic acid switches and sensors**. RNAs have been found whose activities are modulated either by proteins or by small-molecule metabolites, and both kinds of system are described. Finally, the notion of exploiting naturally occurring RNA switches for drug development is discussed. Overall, the studies described in this book show that interesting and useful molecular-scale switches and sensors can be made out of nucleic acids, by both artificial and natural means.

Scott K. Silverman, Ph.D.

SECTION I
Artificial Nucleic Acid Switches and Sensors

Allosteric Ribozymes as Molecular Switches and Sensors

Garrett A. Soukup*

Abstract

Since the discovery of RNA catalysts, biotechnology has focused heavily on utilizing ribozymes as reagents to control RNA processing and gene expression. However, ribozymes can also be manipulated to report events that affect their folding and catalysis. As with protein enzymes, ribozyme activity is dependent upon the ability of the biopolymer to form secondary and tertiary structures that establish the active conformation. Molecular engineering efforts have exploited the structure-function relationship of RNA catalysts to create novel allosteric ribozymes whose activities are modulated by the binding of specific effector molecules. Such efforts are facilitated by the diversity of RNA-ligand interactions, the general predictability of nucleic acid folding, and the relative ease of in vitro RNA synthesis and manipulation. Engineered allosteric ribozymes are inherently molecular sensors for their cognate ligands, and they function as molecular switches regulated by ligand interaction. Consequently, allosteric ribozymes are finding utility in various molecular sensor applications and as genetic regulatory switches.

Introduction

RNA is a highly versatile nucleic acid biopolymer that fulfills numerous roles in biology, including information transfer, protein synthesis, and RNA processing. Among these biological functions, one of the most intriguing is the ability of RNA to catalyze biochemical reactions (Table 1)[1-15] that are central to peptide bond formation and RNA cleavage or splicing.[16,17] RNA catalysts, like their protein enzyme counterparts, are endowed with their catalytic properties by the ability of the biopolymer to form intricate secondary and tertiary structures that position functional groups at an active site. Although natural RNA catalysts (ribozymes) are limited in regard to the number of functional motifs that have been identified, a variety of artificial catalysts have been generated in vitro that demonstrate the true versatility of nucleic acid biopolymers as catalysts of biochemical reactions.[18,19]

Aside from catalytic potential, RNA is a structurally dynamic biopolymer that can exhibit conformational transitions dependent upon its environment and, in particular, upon interactions with ligands. In many cases, ligand-induced conformational changes are an integral aspect of biological RNA function, effecting processes such as transcription termination and ribosomal translation.[20-26] Moreover, RNA is extremely adept with regard to molecular recognition and discrimination of a variety of ligands. A multitude of in vitro-generated ligand-binding

*Garrett A. Soukup—Department of Biomedical Sciences, Creighton University School of Medicine, 2500 California Plaza, Omaha, Nebraska 68178, U.S.A.
Email: gasoukup@creighton.edu

Nucleic Acid Switches and Sensors, edited by Scott K. Silverman. ©2006 Landes Bioscience and Springer Science+Business Media.

Table 1. Natural RNA catalysts

Ribozyme	Activity	Reference
Hammerhead	Self-cleavage	1,2
Hairpin	Self-cleavage	3
HDV	Self-cleavage	4
VS	Self-cleavage	5
glmS	Self-cleavage	6
β-globin	Self-cleavage	7
RNase P	tRNA cleavage	8
Group I intron	Self-splicing	9
Group II intron	Self-splicing	10,11
Spliceosome	RNA splicing	12,13
Ribosome	Peptidyl transfer	14,15

RNAs, or aptamers, demonstrate that RNA can specifically bind molecules ranging from small compounds to proteins.[27-29] Additionally, aptamers typically exhibit adaptive binding or conformational transition upon ligand interaction.[30] Nowhere has the influence of ligand-induced conformational changes on biological RNA structure and function been better appreciated than in the recent discovery of riboswitches (see Chapter 6).[31,32] Riboswitches contain natural aptamers that bind metabolic compounds including cofactors,[33-38] amino acids,[39-41] purine bases,[42,43] and an aminosugar.[6] Resident largely within the 5' untranslated regions of prokaryotic messenger RNAs, riboswitches modulate gene expression through metabolite-induced conformational changes that effect transcription termination, translation initiation, or RNA processing.[6,44,45]

RNA catalysis and the structural dynamics of RNA-ligand interactions form the basis for an artificial class of RNA activities termed allosteric ribozymes.[46-49] Molecular engineering strategies have enabled the generation of allosteric ribozymes by integrating ligand-binding and catalytic functionalities in ways that achieve effector-dependent conformational changes that modulate ribozyme activity and establish either allosteric activation or inhibition (Fig. 1). Such efforts have been facilitated by the diversity of known aptamer and ribozyme motifs, the

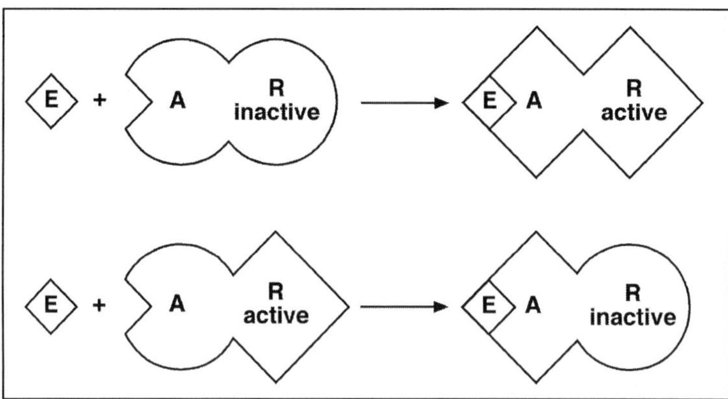

Figure 1. Modulation of ribozyme activity through allosteric activation (top) or inhibition (bottom). Ribozyme (R) and aptamer (A) domains are integrated such that effector (E) binding supports or disrupts the catalytically active conformation.

general predictability of nucleic acid folding, and the relative ease of RNA synthesis and manipulation using in vitro techniques. Since the activity of an allosteric catalyst is dependent upon interaction with the effector molecule, allosteric ribozymes can serve as either molecular sensors for their ligands or as molecular switches that are regulated by effector binding. Consequently, allosteric ribozymes are finding utility in a variety of applications for molecular detection and as genetic regulatory switches. The goals of this chapter are to convey general concepts and techniques used to develop allosteric ribozymes, and to consider the various applications of allosteric ribozymes as tools for exploring and manipulating biology.

Genesis of Allosteric Ribozymes

A variety of allosteric ribozymes have been engineered to respond to effector molecules that include metal ions, biological metabolites, pharmaceutical agents, peptides and proteins, and oligonucleotides (Table 2; see also Chapters 2 and 3). Such catalysts are engineered by one or more techniques that include modular rational design and combinatorial selection strategies for functionally integrating ligand-binding and catalytic activities. These strategies have been applied successfully to a number of naturally occurring or artificial ribozymes that perform RNA cleavage, splicing, or ligation, and also DNA catalysts (deoxyribozymes) that perform RNA cleavage or DNA ligation. Despite the catalytic platform utilized, a prevailing theme in the derivation of allosteric ribozymes is that integration of ligand-binding and catalytic domains occurs at requisite features of secondary structure to create interdependency between the functional domains.

Modular Rational Design Strategies

Modular rational design seeks to integrate known ligand-binding and catalytic RNA domains through defined and predictable structural elements. Aptamers are key components in the design of allosteric catalysts, where adaptive binding of ligand provides the driving force for conformational changes that ultimately influence ribozyme activity. In particular, aptamers that bind adenosine 5'-triphosphate (ATP),[88] flavin mononucleotide (FMN),[89] and theophylline[90] have been frequently utilized. These aptamers have especially benefited modular rational design strategies because their precise structures and sites of ligand interaction have been determined by NMR spectroscopy.[91-93] Similarly, the self-cleaving hammerhead ribozyme has been widely exploited in the design of allosteric ribozymes, as its versatile three-stem structure which organizes a well-characterized and compact catalytic core has been solved by X-ray crystallography.[94-101] Thus, biochemical and structural studies have revealed the structure-function relationships of both aptamer and catalytic RNAs and emboldened their utility as components in modular rational design strategies for constructing allosteric ribozymes.

Table 2. Allosteric ribozymes

Genesis	Effector Class	References
Modular rational design	Biological metabolites	50-60
	Pharmaceutical agents	50,50-58,61
	Peptides and proteins	62-65
	Oligonucleotides	63,66-75
Combinatorial selection	Metal ions	76,77
	Biological metabolites	54,56,58,78-80
	Pharmaceutical agents	58,81-84
	Peptides and proteins	85,88
	Oligonucleotides	53,87

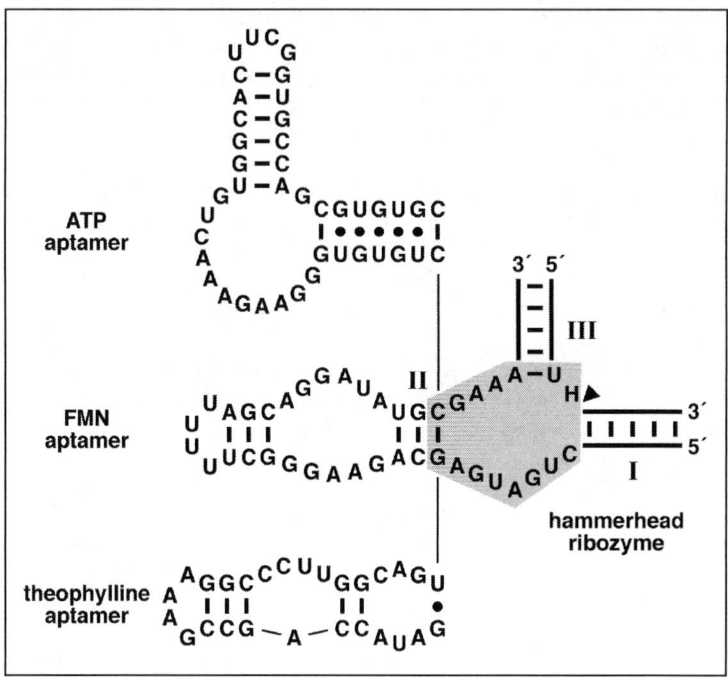

Figure 2. Rational design of allosteric ribozymes activated by ligand-dependent structure stabilization. Each aptamer is integrated with the hammerhead ribozyme through relatively short or weak stem II elements that are stabilized by ligand binding. The adjacent core of the hammerhead ribozyme is shaded, and the arrowhead indicates the cleavage site. H represents A, C, or U, and dashes indicate base pairing.

A principal means of integrating aptamers and catalytic RNAs is through a shared element of secondary structure that is required for the organization of both functional domains. However, the shared element is typically minimized or designed such that it is thermodynamically weak in relation to a base-paired structure that would support independent activity of either functional domain. The resulting construct is thus crippled in the sense that it is poorly organized in the absence of ligand and less apt to perform catalysis. Ligand binding, however, promotes folding of the aptamer domain, stabilizes the shared element of secondary structure, and prompts folding and catalysis within the ribozyme domain. Consequently, allosteric activation is mediated by ligand-dependent stabilization of the catalytically competent structure. Early examples of such allosteric ribozymes include ATP-,[50] FMN-,[52,55] and theophylline-dependent[52,55] self-cleaving hammerhead ribozymes that incorporate the respective aptamer sequences (Fig. 2). In these designs, the ligand binding site and the catalytic core are closely juxtaposed through a shared element of one or more base pairs, often incorporating thermodynamically less stable G•U wobble pairs. Allosteric ribozymes designed in this manner typically exhibit rate constants in the presence of ligand that are one to two orders of magnitude greater than those determined in the absence of ligand. Allosteric activation of ribozymes is therefore comparable to that observed for many protein enzymes,[102-103] suggesting that allosteric regulation of ribozyme activity could in principle exert significant effects on biological processes.

An important consideration in modular rational design is the effect of the integrated construct on ligand-binding and catalytic activities relative to the activities of the individual components. Inherent to the mechanism of allosteric activation involving structure stabilization are adverse effects on general RNA folding and function. With respect to catalysis, allosteric ribozyme activities often approach but rarely reach the maximum rate constant observed for the analogous unmodified ribozymes. For example, the hammerhead ribozyme performs self-cleavage with a rate constant of ~1 min^{-1}.[97] However, modular rational design affords FMN-dependent ribozymes with maximal rate constants that are less than 20% that of the unmodified ribozyme.[52,55] Additionally, kinetic analyses of such allosteric ribozymes are biphasic, indicating that a fraction of the RNA is misfolded and slow to respond to ligand. With respect to ligand binding, inherent disorganization of the allosteric ribozyme can reduce the affinity of the aptamer domain for its cognate ligand. For example, FMN-dependent ribozymes exhibit apparent dissociation constants (K_d values) that are 10-fold or 260-fold greater than that of the independent FMN-binding aptamer.[52,55] While these aspects of allosteric ribozyme performance are not easily overcome by rational design principles, they can be optimized by combinatorial selection strategies, which are addressed in the following section. However, the rational design studies demonstrate that allosteric ribozymes retain the molecular recognition and discrimination capabilities of their aptamer components, but provide the benefit of catalytic output to report the binding event. Consequently, modular rational design is a facile route for the construction of allosteric ribozymes as molecular sensors.

A mechanistically distinct mode of ligand recognition and allosteric regulation of catalysis is represented by oligonucleotide-dependent ribozymes. As ligand recognition is simply mediated by Watson-Crick base-pairing, the development of oligonucleotide-dependent ribozymes is highly amenable to rational design. A variety of strategies for achieving oligonucleotide-dependent catalysis have been demonstrated utilizing the hammerhead ribozyme (Fig. 3). In one strategy, oligonucleotide hybridization to allosteric ribozymes serves to organize the catalytic core and activate catalysis (Fig. 3A).[66,67,70] A different strategy achieves oligonucleotide-dependent activation by competing an alternative structure that attenuates ribozyme activity (Fig. 3B).[71,74,75] In such "TRAP" (targeted ribozyme-attenuated probe) designs, oligonucleotide hybridization sequesters an attenuator sequence that otherwise inactivates the ribozyme by directly binding and disrupting the catalytic core. In a third strategy that requires an expanded definition of allostery, oligonucleotide hybridization effects ribozyme activity by requisite participation in substrate binding (Fig. 3C).[72,73] Such strategies illustrate the amenability of nucleic acid catalysts to oligonucleotide regulation and the applicability of allosteric ribozymes to nucleic acid sequence detection. Moreover, aspects of these strategies can be used in conjunction with aptamer-based designs to facilitate ligand-dependent ribozyme catalysis.[65,75]

Other strategies for modular rational design require and incorporate knowledge of the tertiary structure of RNA aptamers and catalysts to achieve customized mechanisms of allosteric regulation. For example, an ATP-dependent hammerhead ribozyme specifically exploits the conformational rigidity and tertiary structure of an ATP-bound aptamer domain to strategically position a helical segment that sterically hinders formation of the ribozyme domain's active conformation (Fig. 4A).[50,51] However, conformational flexibility of the aptamer domain in the absence of ligand permits ribozyme catalysis. This manner of ATP-dependent allosteric inhibition has enabled an examination of core sequence fitness in hammerhead ribozyme catalysis by maintaining ribozyme inactivity during preparation.[104] Another design strategy inspired by the tertiary requirements of RNA folding and catalysis has utilized polypeptide-RNA interactions to functionally replace a direct RNA contact required for *Tetrahymena* group I intron self-splicing activity (Fig. 4B).[62] By substituting RNA tetraloop and tetraloop receptor elements with RNA binding sites for specific polypeptides, protein-dependent self-splicing introns have been developed that function both in vitro and in vivo.[86] Such allosteric ribozymes provide downstream opportunities for monitoring and exploring RNA-protein interactions.

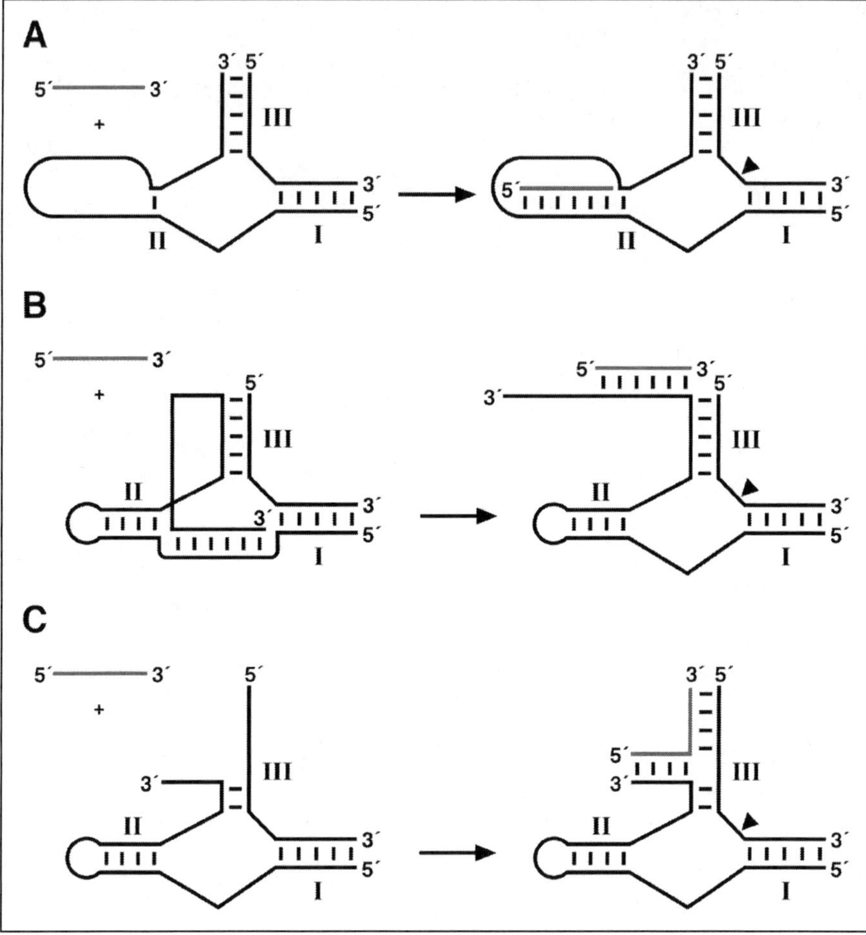

Figure 3. Rational design of oligonucleotide-dependent ribozymes. A) Structure stabilization. Oligonucleotide hybridization promotes organization of adjacent stem structure and therefore promotes catalysis. B) Targeted ribozyme-attenuated probe (TRAP) design. Oligonucleotide hybridization sequesters an attenuator sequence and alleviates attenuator disruption of the catalytic core. C) Expansive allostery. Oligonucleotide hybridization participates in substrate interaction and promotes catalysis. Arrowheads at the cleavage site indicate active ribozymes.

Combinatorial Selection Strategies

While modular rational design has laid the foundation for integrating aptamer and ribozyme domains to achieve allosteric regulation of catalysis, the application of combinatorial selection strategies is immensely useful for optimizing allosteric ribozyme performance and for generating novel effector-dependent activities. Combinatorial strategies empower the development of allosteric ribozymes by resolving functional sequences from random-sequence populations of prospective allosteric catalysts using in vitro selection techniques.[27,28] Depending upon population design (Fig. 5), selection for allosteric ribozyme activity ("allosteric selection") has facilitated the isolation of structural elements that better integrate existing aptamer and ribozyme

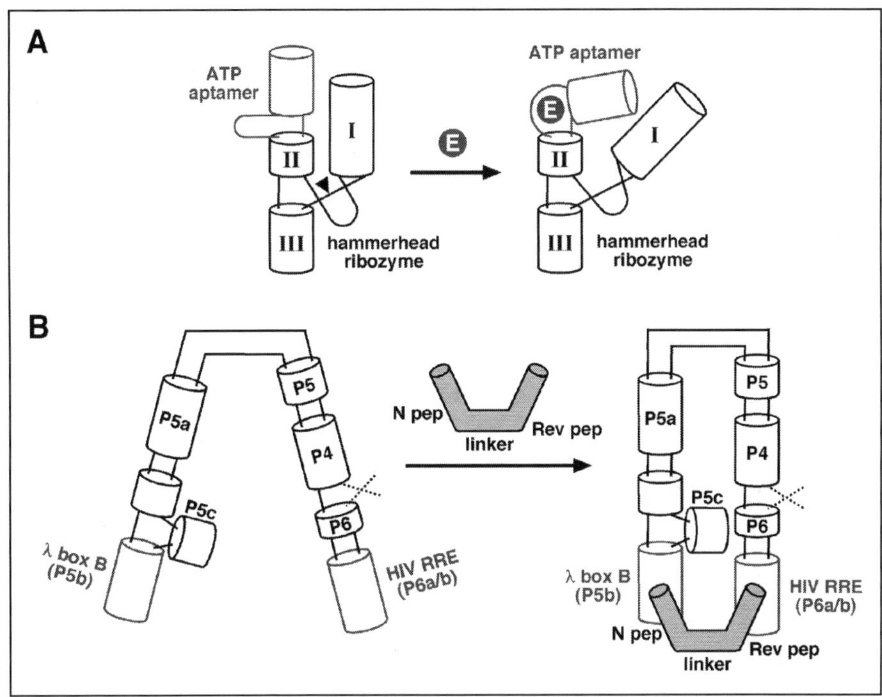

Figure 4. Rational design of allosteric ribozymes based on tertiary structure. A) ATP-dependent allosteric inhibition of hammerhead ribozyme activity. Steric hindrance between helical segments (cylinders) within the aptamer and ribozyme domains prevents ribozyme activity (arrowhead) upon effector (E) binding. B) Polypeptide-dependent allosteric activation of the *Tetrahymena* group I intron. Polypeptide-binding RNA domains from bacteriophage λ and human immunodeficiency virus (HIV) replace the P5b and P6a/b stems which form a required RNA-RNA interaction. Fusion of the λ box B-binding N peptide (pep) to HIV Rev-responsive element (RRE)-binding Rev peptide through a linker produces a polypeptide effector (shaded cylinder) capable of promoting proper RNA folding and self-splicing activity. Only P4-P6 of the *Tetrahymena* intron is depicted. Dashed lines indicate a connection to the remainder of the ribozyme structure.

domains (Fig. 5A), the isolation of novel ligand-binding domains (Fig. 5B), and the isolation of variant ligand-binding domains with altered effector specificity or affinity (Fig. 5C).

Allosteric selection seeks to isolate from populations of potential catalysts those individuals exhibiting allosteric activation or inhibition through an iterative process of selection and amplification (Fig. 6A). Populations of potential catalysts are typically derived by transcription from synthetic DNA templates containing randomized sequence segments. Such populations are expected to contain a continuum of catalytic activities ranging from nonfunctional to ligand-independent, and from ligand-activated to ligand-inhibited (Fig. 6B). The challenge of allosteric selection is to partition these activities through a two-step selection process that favors the isolation of individuals that exhibit the desired activity. For example, to isolate individuals that exhibit allosteric activation, the population is first purged of members that exhibit activity in the absence of ligand by isolating the inactive fraction following a preselection reaction lacking ligand. Subsequently, the remaining population is enriched for members that exhibit activity in the presence of ligand by isolating the active fraction following a selection reaction including ligand. Reverse transcription, PCR amplification, and transcription yield a

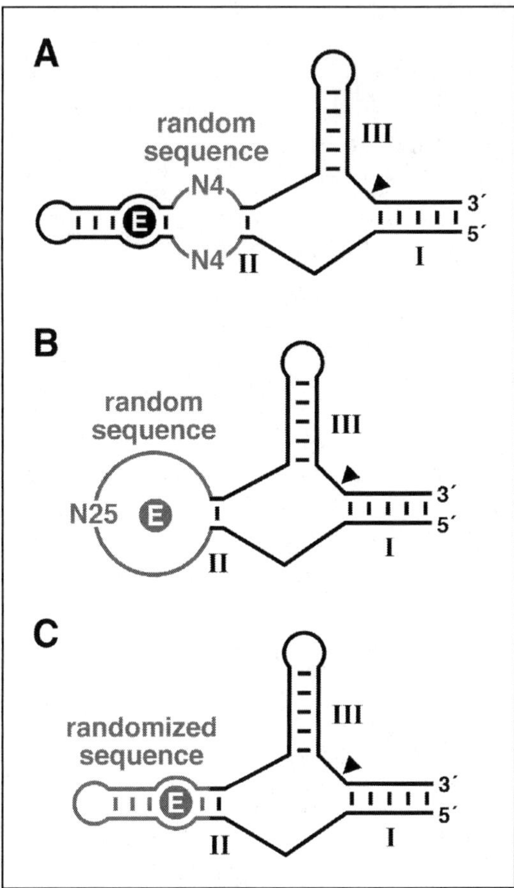

Figure 5. Random-sequence population designs for selection of allosteric ribozymes. A) Population for selection of communication modules. Short random-sequence segments (e.g., N4 representing four random-sequence positions) are used to join existing aptamer and ribozyme domains. B) Population for selection of novel ligand-binding domains. A larger random sequence segment (e.g., N25) is appended to a ribozyme domain. C) Population for selection of aptamers with altered specificity or affinity. Each nucleotide position within the aptamer domain is partially randomized (mutagenized) to create a degenerate population.

subsequent population of catalysts that have completed one cycle of selection and amplification. Populations are iteratively processed and monitored throughout the course of allosteric selection to obtain those individual sequences that best meet the selection criteria for ligand-activated function. Catalysts that exhibit allosteric inhibition can be identified using a reciprocal approach including ligand in preselection reactions and excluding ligand from selection reactions.

Although modular rational design has served to identify means of integrating aptamer and ribozyme domains through shared elements of secondary structure, allosteric selection can identify such elements with little preconception regarding precise sequence, secondary structure, and mechanistic function. By randomizing short 4-5 base pair segments that join aptamer and

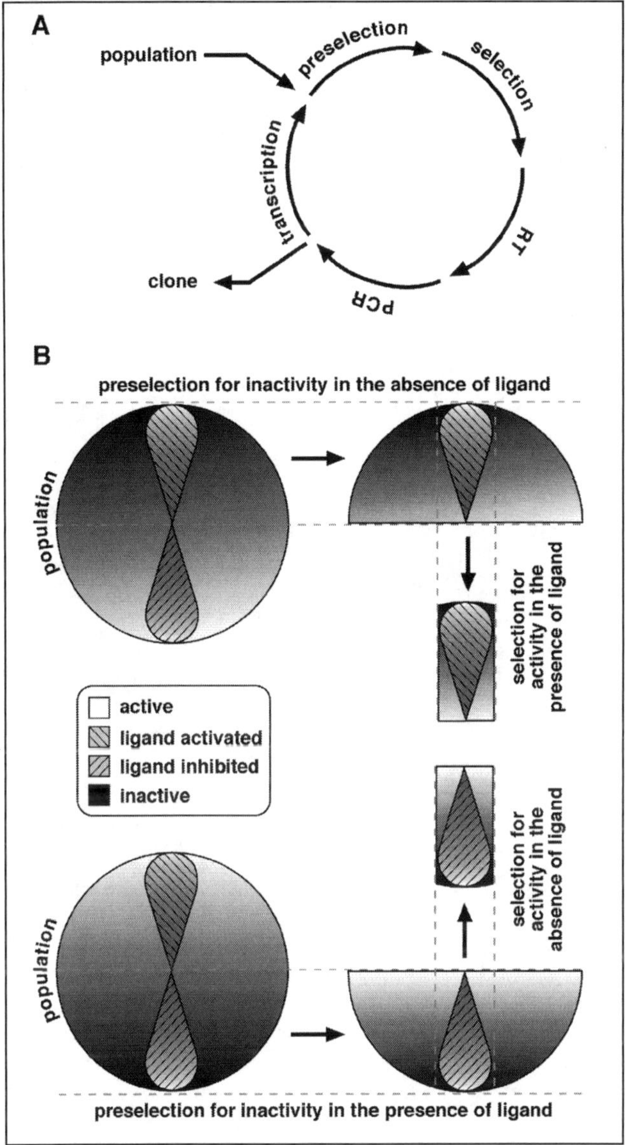

Figure 6. In vitro selection of allosteric ribozymes. A) Selection scheme. Allosteric catalysts are isolated through iteration of a selection cycle in which the population is subjected to a two-step selection process, and the selected fraction is amplified and regenerated by reverse transcription (RT), polymerase chain reaction (PCR), and transcription of RT-PCR products. B) Allosteric selection. The two-step selection process seeks to isolate allosteric catalysts by partitioning the population through preselection and selection reactions based on activity in the absence or presence of ligand. Allosteric catalysts that are activated or inhibited by ligand are derived by reciprocal approaches (top or bottom) that exclude ligand in preselection reactions and include ligand in selection reactions, or include ligand in preselection reactions and exclude ligand in selection reactions.

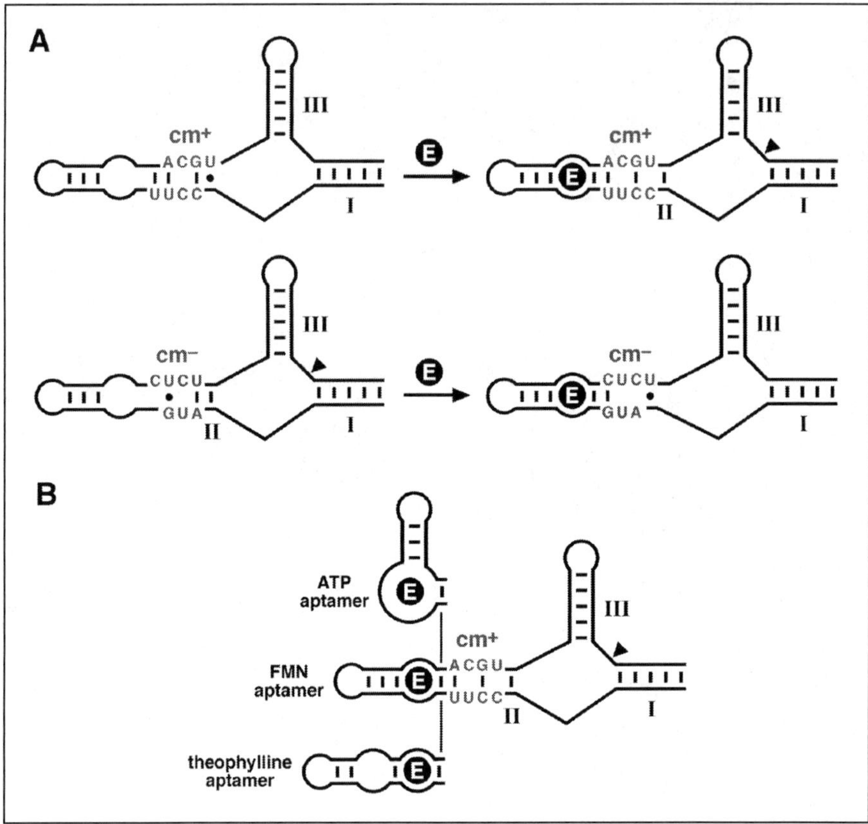

Figure 7. Communication module function and utility. A) Communication modules for allosteric activation (cm⁺) or inhibition (cm⁻). Communication modules are modeled to adopt alternate secondary structures dependent upon effector (E) binding that either activate or inhibit ribozyme. Arrowheads indicate active ribozymes. B) Incorporation of communication modules in rational design. Communication modules can enable aptamer domain exchange and a generally adaptive mode of allosteric regulation.

ribozyme domains (Fig. 5A), relatively small populations of prospective allosteric catalysts can be generated that include all possible sequence combinations. Allosteric selection from such populations serves to identify unique elements, or communication modules, that provide for allosteric function. For example, communication modules that provide for either FMN-dependent allosteric activation or inhibition of hammerhead ribozyme self-cleavage have been identified,[54] where communication module function is modeled to adopt alternative secondary structures in the absence or presence of ligand that are either compatible or incompatible with ribozyme structure and function (Fig. 7A). Moreover, communication modules can facilitate modular rational design strategies by serving in a generalizable manner to functionally integrate different ligand-binding domain. For example, ATP- and theophylline-dependent ribozymes are readily designed by joining the appropriate aptamer domain to the hammerhead ribozyme through a FMN-derived communication module that enables allosteric activation (Fig. 7B).[54]

Allosteric selection for communication module function has been applied toward the development of allosteric hammerhead,[54,80,81] hepatitis delta virus (HDV),[58] and artificial ligase ribozymes[56] utilizing FMN-, theophylline-, and adenosine 5'-diphosphate (ADP)-binding aptamer domains. In general, such allosteric ribozymes exhibit better allosteric performance than those obtained by modular rational design alone. Where selective pressure has been applied to the fullest, theophylline-dependent hammerhead ribozymes have been isolated that possess rate constants that equal that of the unmodified hammerhead ribozyme, and that exhibit more than three orders of magnitude of allosteric activation.[81] Moreover, a number of derivative communication modules exhibit the capacity to functionally integrate different domains through modular rational design.[54,56,58,81] The most versatile module, derived through allosteric selection for theophylline-dependent HDV ribozymes, has facilitated the design of ATP-, FMN-, and theophylline-dependent allosteric ribozymes that include the HDV, hammerhead, artificial X-motif, and *Tetrahymena* group I ribozymes.[58] Therefore, allosteric selection can solve the challenge of functionally integrating aptamer and catalytic RNAs and provide novel functional domains that enhance the downstream development of allosteric ribozymes.

A more challenging application of allosteric selection includes the goal of developing novel ligand-binding domains from random sequence. Rather than appending a specified aptamer domain, a large random-sequence segment of 25 to 40 nucleotides is joined to a catalytic RNA domain (Fig. 5B). In this manner, relatively large populations of prospective allosteric catalysts are conceivable (10^{15} to 10^{24} possible sequence combinations), of which 10^{14} to 10^{15} sequences are feasibly sampled in a particular in vitro selection experiment. As aptamers for virtually any ligand can be generated from similarly sized random-sequence RNA populations,[27,28] such populations of potential allosteric catalysts are predicted to include individuals that are allosterically regulated by a variety of effector molecule types. Allosteric selection is made challenging by the fact that the desired activity is relatively rare in the population. Consequently, stringent selection conditions are required to reduce the isolation of "selfish" molecules that meet selection criteria by other than the intended means.[78] Demonstrations of this strategy for developing novel effector specificities include the isolation of allosteric hammerhead ribozymes that are activated by 2',3'-cyclic nucleotide monophosphate (cNMP) compounds including cAMP, cGMP, and cCMP,[78] quinolone derivatives including doxycycline and pefloxacin,[82,83] caffeine and aspartame,[84] and a variety of divalent metal ions.[77] For such compounds, allosteric selection provides a distinct advantage for isolating ligand-binding RNAs. Traditional aptamer selections typically require that ligand is immobilized on a solid support to facilitate chromatographic separation of bound and unbound RNAs, thereby presenting a steric impediment to molecular recognition. Allosteric selection, on the other hand, utilizes effector free in solution to identify ligand-binding RNAs, thus avoiding steric considerations and also permitting recognition of effectors that cannot be easily immobilized.[77] Allosteric selection has enabled the identification of aspartame-specific RNAs that were not obtainable through traditional aptamer development.[84]

Allosteric selection can similarly be applied to alter the properties of effector-binding domains. By partially randomizing (i.e., mutagenizing) the aptamer domain of existing allosteric ribozymes (Fig. 5C), variant ligand-binding domains with altered effector specificity or affinity can been generated. For example, allosteric selection has enabled the identification of nucleotide substitutions within theophylline- or cGMP-binding domains that respectively favor 3-methylxanthine recognition or effect an order of magnitude improvement in cGMP affinity.[79,81] Interestingly, allosteric selection techniques have been applied to both combinatorialized RNA and peptide ligands to identify novel RNA-protein interactions that constitute new effector specificities.[86] Therefore, allosteric selection can draw on a variety of combinatorial strategies to identify novel ligand-binding RNAs and to develop allosteric ribozymes with specified effector-dependent activities.

Allosteric Ribozymes as Molecular Sensors

Any particular allosteric ribozyme is inherently a molecular sensor for its cognate ligand. Because the observed rate constant (k_{obs}) for an allosteric ribozyme responds measurably to effector concentration, allosteric ribozymes not only report the presence of effector but enable quantification of ligand as well. The effective range of quantification, or "dynamic range", is set both by the magnitude of allosteric modulation and the apparent dissociation constant (K_d) for effector interaction. Dynamic range therefore operates over a range of ligand concentration equivalent to the magnitude of allosteric modulation up to the apparent K_d. For example, a cGMP-dependent allosteric hammerhead ribozyme that is activated over three orders of magnitude when saturated with cGMP and that exhibits an apparent K_d of ~2 mM is capable of determining cGMP concentration throughout the μM to mM range.[78] When such operational parameters are congruent with practical considerations, allosteric ribozymes can find utility in a variety of applications as molecular sensors for drug discovery, protein and metabolite analyses, and nucleic acid sequence detection.

While allosteric ribozymes are already effective for reporting and quantifying target ligands, their utility is further enhanced by various detection strategies. Activity has primarily been assessed by quantifying the reaction of radiolabeled allosteric ribozymes or substrates versus time to determine ligand-dependent observed rate constants. However, detection strategies that incorporate fluorophore-labeled substrates[105,106] have enabled real-time monitoring of allosteric ribozyme activity to expedite analysis and facilitate high-throughput applications (Fig. 8).[63,65,75,80,84] For example, fluorophore-labeled substrates for hammerhead ribozyme cleavage can either incorporate[105] or be positioned adjacent to a quencher dye[80] such that

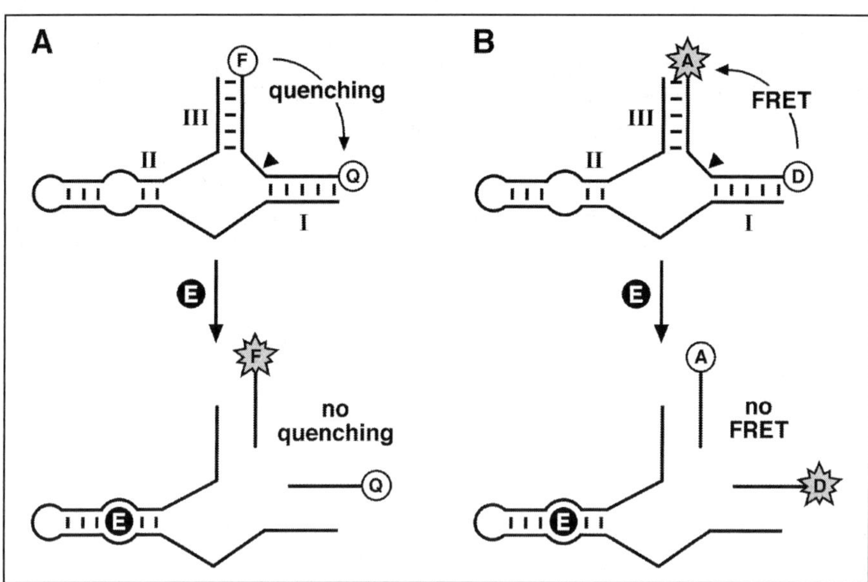

Figure 8. Fluorescence strategies for real-time analysis of allosteric ribozyme activity. A) Fluorescence quenching. Ribozyme substrate incorporating both fluorophore (F) and quencher (Q) dyes generate fluorescence signal (star) upon effector (E)-dependent cleavage and product dissociation. B) Fluorescence transfer. Ribozyme substrate incorporating coupled donor (D) and acceptor (A) fluorophores differentially generate acceptor or donor fluorescence (stars) dependent upon cleavage, product dissociation, and change of fluorescence resonance energy transfer (FRET).

fluorescence signal is generated upon substrate cleavage and product release. Alternatively, a dual-labeled substrate containing a coupled donor-acceptor fluorophore pair can be used to monitor the change of fluorescence resonance energy transfer (FRET) signal upon substrate cleavage and product release.[106] The former strategy has been more widely implemented in monitoring allosteric ribozyme activity in molecular sensor applications. See Chapter 4 for more about RNA- and DNA-based sensors that function with fluorescence detection.

Drug Discovery

A distinctive application of allosteric ribozymes as molecular sensors is demonstrated by their ability to assess intermolecular interactions that constitute pharmaceutically relevant complexes. To identify compounds that interact with human immunodeficiency virus (HIV) Rev protein and potentially inhibit Rev function, Hartig et al[63] designed allosteric hammerhead ribozymes activated or inhibited by Rev peptide binding whose activities are easily monitored by real-time detection. The allosteric ribozymes were used to screen a library of 96 antibiotic compounds to identify those that prevent Rev peptide-dependent catalysis (Fig. 9A). Three compounds that functioned in the screen were indeed demonstrated to bind Rev, and one was subsequently shown to inhibit HIV replication in cell culture. Expanding this concept, allosteric ribozymes that are inhibited by intermolecular hybridization to a thrombin aptamer were used monitor the thrombin-aptamer interaction (Fig. 9B).[63] Through the mutually exclusive formation of complexes, the allosteric ribozyme was proficient in detecting protein and peptide inhibitors of thrombin. Therefore, allosteric ribozymes can be adept tools for monitoring intermolecular protein interactions as applied to drug discovery.

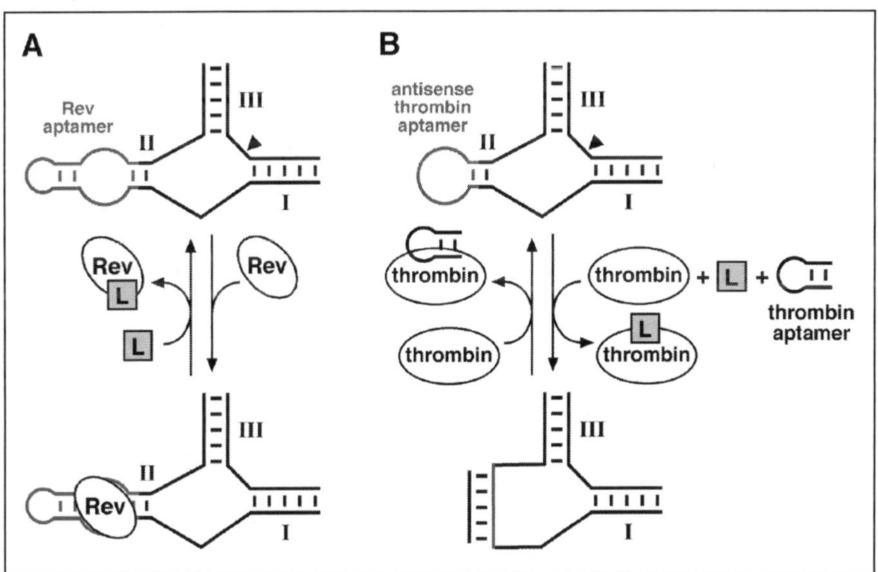

Figure 9. Application of allosteric ribozymes as molecular sensors for drug discovery. A) Molecular sensor for Rev-binding ligands. Allosteric activation of a Rev-dependent hammerhead ribozyme is inhibited by a Rev-binding ligand (L). B) Molecular sensor for thrombin interactions. An allosteric hammerhead ribozyme is inhibited by hybridization to thrombin aptamer and thus senses the mutually exclusive interactions of thrombin with the aptamer or a thrombin-binding ligand that displaces the aptamer. Arrowheads indicate active ribozymes.

Protein and Metabolite Analyses

Many recent examples highlight the utility of allosteric ribozymes as molecular sensors for protein or metabolite analysis. Vaish et al[65] have employed allosteric ribozymes for the detection of the mitogen-activated protein kinase ERK2, which is associated with malignant cell growth and is phosphorylated in its active form. Utilizing aptamers that distinguish unphosphorylated and phosphorylated ERK2, allosteric hammerhead ribozymes were developed that detect and report the activation state of the protein (Fig. 10A). Moreover, the allosteric ribozymes discriminate ERK2 from other closely related protein kinases and also function to detect exogenous ERK2 in complex mixtures that include cell lysate. Srinivasan et al[80] have developed ADP-specific aptamer and allosteric ribozyme sensors for assessing

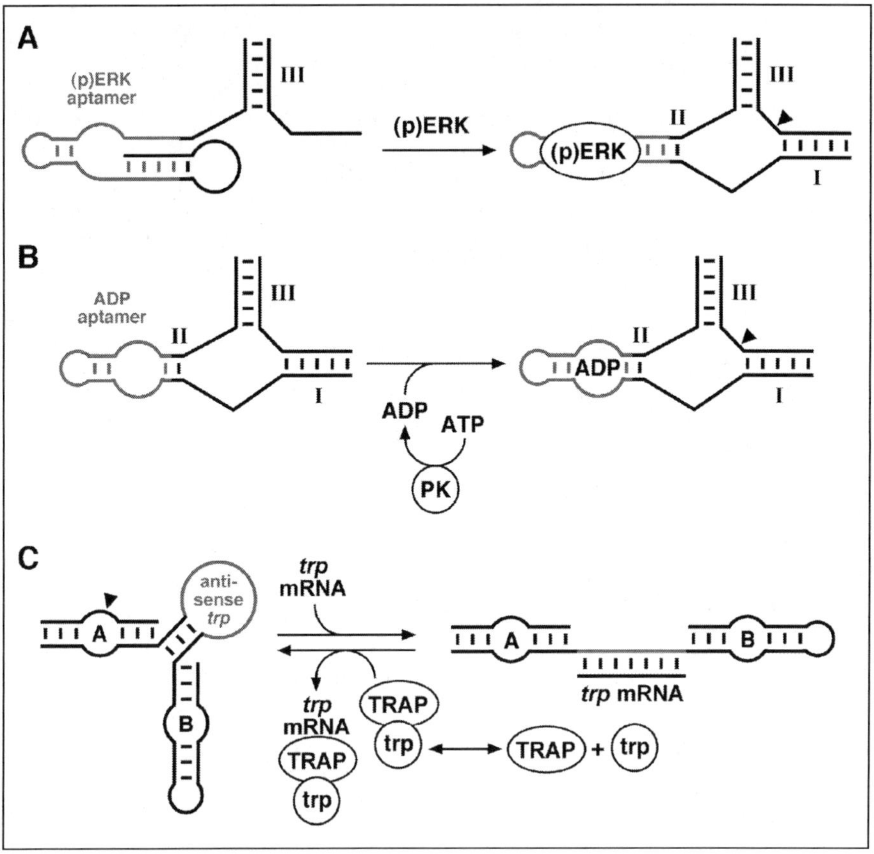

Figure 10. Application of allosteric ribozymes as molecular sensors for protein or metabolite analysis. A) ERK2 molecular sensors. Integration of aptamers specific for unphosphorylated or phosphorylated forms of ERK2 [(p)ERK] with the hammerhead ribozyme produce allosteric catalysts capable of detecting post-translational modification of the protein. B) ADP molecular sensor. An allosteric hammerhead ribozyme activated specifically by ADP is capable of detecting protein kinase (PK) activity and thus protein kinase inhibitors. C) Tryptophan-activated *trp*-RNA-binding attenuation protein (TRAP) sensor. An allosteric hairpin ribozyme is inhibited by hybridization to available *trp* mRNA and thus senses tryptophan-activated TRAP binding and sequestration of *trp* RNA.

protein kinase activity in vitro. While the aptamer and allosteric ribozyme-based assays can be universally applied to detect protein kinase conversion of ATP to ADP, the allosteric ribozyme was demonstrated to detect ERK2 activity and identify a known protein kinase inhibitor in a screen of 77 compounds (Fig. 10B). These examples demonstrate the effectiveness of allosteric ribozymes in discriminating biologically relevant protein targets and their metabolic products in applications that promise to streamline analysis and facilitate drug discovery.

In another example, Najafi-Shoushtari et al[75] have generated allosteric hairpin ribozymes that are either activated or inhibited by a metabolically important regulator of tryptophan biosysnthesis in prokaryotes. Although the allosteric ribozymes are directly regulated by intermolecular hybridization with the leader sequence of *trp* mRNA, they are capable of evaluating tryptophan-activated *trp*-RNA-binding attenuation protein (TRAP) activity in vitro by the mutually exclusive formation of TRAP-mRNA complexes (Fig. 10C). The allosteric ribozymes have thus served to assess the levels of tryptophan and TRAP required to bind the leader sequence of *trp* mRNA, and have demonstrated the ability of allosteric ribozymes to gauge the metabolic parameters of complex regulatory networks.

The biotechnological utility of allosteric ribozymes as molecular sensors is further impacted by demonstrations that allosteric catalysts can be immobilized and arrayed in addressable microplate formats.[76,107] Seetharaman et al[76] have derivatized and immobilized various metabolite-dependent allosteric hammerhead ribozymes on gold to produce a biosensor array that is capable of discriminating metabolites alone or in complex mixtures through effector-dependent ribozyme cleavage and loss of radiolabeled product (Fig. 11A). The biosensor array was demonstrated to properly assess the phenotype and genotype of bacterial strains with regard to adenylate cyclase expression by quantitatively detecting cyclic AMP in samples of culture media. Hesselberth et al[107] have exploited allosteric ligase ribozymes that acquire an affinity tag to produce a microplate array of metabolite- and protein-dependent ribozymes (Fig. 11B). Through effector-dependent acquisition of a biotinylated substrate, radiolabeled allosteric ligase ribozymes are immobilized to streptavidin-coated plates for analysis. Such arrays were similarly demonstrated to detect and quantify effectors alone or in complex mixtures with a level of sensitivity that exceeds typical immunosorbent assays. Therefore, allosteric ribozymes can be valuable components of biosensor arrays designed to qualitatively and quantitatively assess protein and metabolite targets of interest.

Nucleic Acid Sequence Detection

As described earlier, ribozymes are highly tractable agents for modular rational design of oligonucleotide-dependent catalysts. In various strategies of allosteric regulation, either RNA or DNA can function as an effector by aiding substrate recognition,[72,73] activating attenuated ribozymes,[71] or influencing ribozyme activity in more direct fashion.[66,67,70] Although oligonucleotide-dependent ribozymes will not replace established methodologies for in vitro nucleic acid sequence detection such as quantitative PCR, there are likely to be certain applications for which they are well-suited reagents. In one example, Hartig et al[74] have designed and utilized oligonucleotide-dependent hairpin ribozymes to detect microRNAs, which are biologically important small RNAs that are difficult to assess by standard methodologies such as PCR and northern blot analysis. While allosteric ribozyme detection of microRNAs was demonstrated to exceed the sensitivity of molecular beacons, it is presently unclear how such ribozymes might compare in sensitivity to other enzyme-linked methodologies for signal amplification and detection. The intracellular application of oligonucleotide-dependent allosteric ribozymes holds the promise of enabling detection of target viral or host sequences.

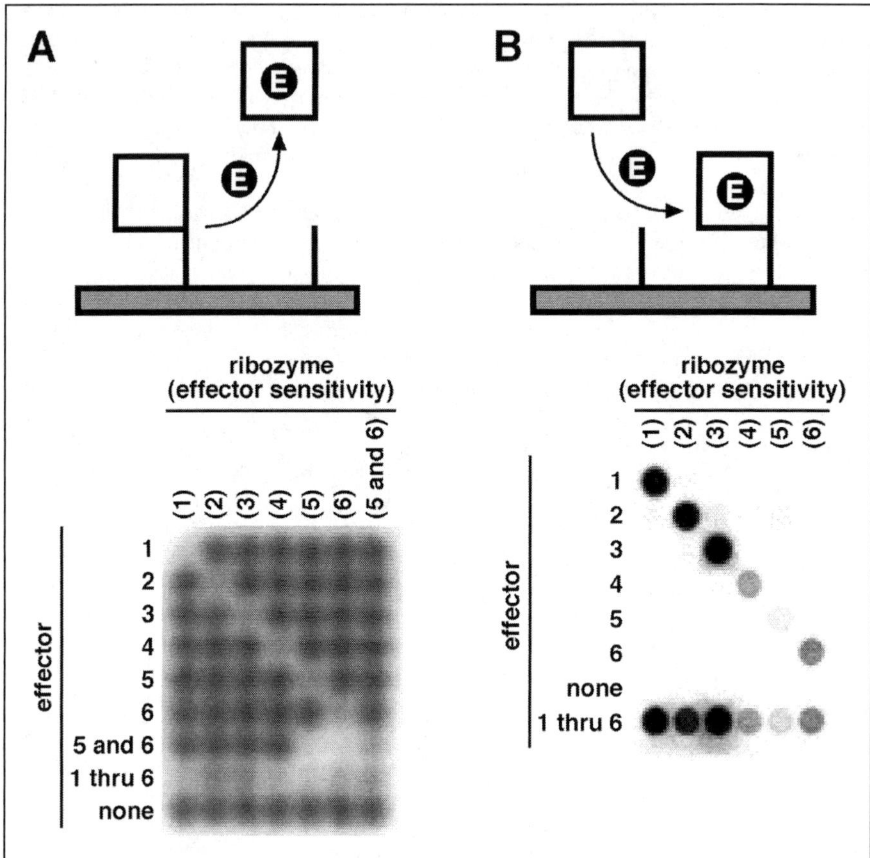

Figure 11. Application of allosteric ribozymes in biosensor arrays. A) Microplate array for self-cleaving ribozymes. Labeled allosteric hammerhead ribozymes immobilized on a solid surface are released upon effector (E)-activated self-cleavage. An array depicting the response of various metabolite-dependent ribozymes (columns) to metabolite-containing solutions (rows) is depicted, where loss of signal indicates activity. Adapted from reference 76. B) Microplate array for ligase ribozymes. Labeled allosteric ligase ribozymes are immobilized on a solid surface upon effector (E)-activated ligation to substrate. An array depicting the response of various metabolite- and protein-dependent ribozymes (columns) to metabolite- and/or protein-containing solutions (rows) is depicted, where gain of signal indicates activity. Reproduced from: Hesselberth JR et al, Anal Biochem 312:106-112,[107] ©2003, with permission from Elsevier.

Allosteric Ribozymes as Genetic Regulatory Switches

Ribozymes have long been the focus of biomedical and biotechnological efforts to control RNA processing and gene expression.[108] Allosteric ribozymes afford a means of imparting regulatory control over intracellular ribozyme functions. Towards this goal, allosteric ribozymes have been demonstrated to function as genetic regulatory switches that control RNA degradation or processing in response to oligonucleotide, pharmaceutical, and protein ligands. Consequently, allosteric ribozymes might serve as regulatable therapeutic or biotechnological agents for controlling gene expression.

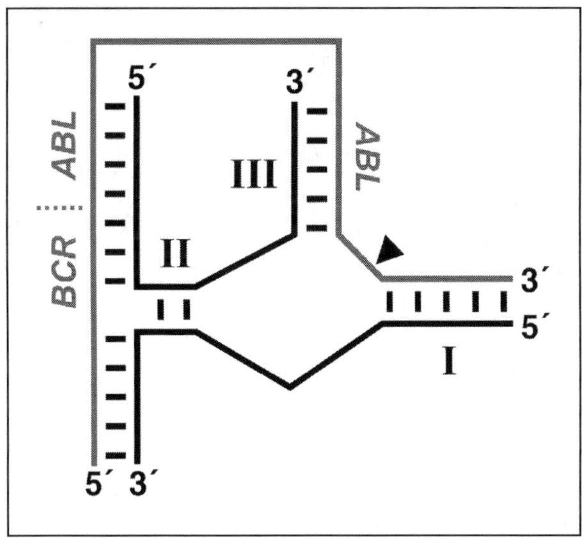

Figure 12. Application of an allosteric hammerhead ribozyme to sequence-specific mRNA detection and cleavage. Depicted is the targeted detection and cleavage of the *BCR-ABL* mRNA by a two-part trans-acting hammerhead ribozyme. The allosteric ribozyme is activated specifically by hybridization to the *BCR-ABL* junction (dashed line) while cleavage (arrowhead) occurs at a different ribozyme-compatible site.

RNA Degradation

In chronic myelogenous leukemia (CML), malignancy is associated with expression of a chimeric *BCR-ABL* gene resulting from a chromosomal translocation. To destroy *BCR-ABL* mRNA and induce apoptosis in CML cells, Taira and coworkers[67,68] have designed oligonucleotide-dependent allosteric hammerhead ribozymes termed maxizymes that detect the *BCR-ABL* junction and specifically cleave the aberrant mRNA in CML cells while leaving normal *ABL* mRNA intact (Fig. 12). The design strategy, which enables detection of a target sequence at one site and cleavage at a different ribozyme-compatible site, has been shown to be generally applicable to other target substrates.[69] Moreover, when such allosteric hammerhead ribozymes are expressed in CML cells or acute lymphoblastic leukemia (ALL) cells, they have been demonstrated to induce apoptosis and to suppress the progression of leukemia in mouse models.[109-111] Consequently, allosteric ribozymes might be useful agents for discriminating aberrant messages and facilitating highly-specific targeted RNA destruction in gene therapy strategies for disease treatment.

RNA Maturation

Self-splicing group I introns are relatively large ribozymes that catalyze intron excision and concomitant ligation of precursor tRNA, rRNA, and mRNA transcripts in numerous prokaryotic and eukaryotic organisms.[112,113] Despite their comparatively complex structure, group I introns are amenable to allosteric regulation using modular rational design principles that have proven effective for simpler catalysts.[58,61,62] Moreover, intracellular function of allosteric group I introns has been demonstrated to regulate gene expression in response to effectors. For example, Thompson et al[61] have developed theophylline-dependent group I introns derived from the bacteriophage T4 *thymidylate synthase* (*td*) gene. Theophylline-dependent *td* intron function and thus *td* mRNA maturation and expression of thymidylate synthase were demonstrated by theophylline-dependent growth of an *E. coli* thymidine auxotroph on minimal media. Employing a single nucleotide mutation that alters the specificity of the ligand-binding domain to favor 3-methylxanthine recognition,[81] allosteric *td* intron function and bacterial growth was readily changed to become 3-methylxanthine-dependent. Importantly, this work demonstrates that allosteric ribozymes can function intracellularly in response to exogenous effectors as do other aptamer-based strategies for controlling gene expression.[114,115] In another example, Atsumi et al[86] have demonstrated that polypeptide-dependent group I introns such as that described earlier (Fig. 4B)[62] function to enable β-galactosidase expression with dependence upon coexpression of an effector polypeptide in *E. coli*. Therefore, allosteric group I introns can serve as artificial genetic regulatory switches that directly impact RNA maturation in response to exogenous or intracellular cues. Conceivably, allosteric group I catalysts could provide a regulatory aspect to trans-splicing strategies for endogenous target RNA repair.[116]

Perspective

RNA possesses considerable capabilities from molecular recognition to catalytic function, and the prevalence of riboswitches and ribozymes throughout biology underscores how critical these properties are to natural systems of genetic regulation and cellular function. From this perspective, the engineering of allosteric ribozymes that seek to combine both molecular recognition and catalysis into a single functional RNA might seem to have been inevitable. However, the development of aptamers and allosteric ribozymes preceded our present-day knowledge of riboswitch function and undoubtedly impacted our expectations for the biological potential of such RNAs. At each turn, it seems that nature has already explored in some manner what scientists have created in the laboratory. For example, consider the recent discoveries of the glucosamine-6-phosphate (GlcN6P)-dependent ribozyme[6] and the guanosine-dependent β-globin ribozyme.[7] Although it remains to be determined precisely whether such metabolites function as allosteric effectors or coenzymes in their respective mechanisms of ribozyme activation, it appears that biology has already exploited ligand-dependent ribozyme activity as a means of genetic regulation.[6] More to the point, it is evident that our exploration of biology can benefit from insights gained through molecular engineering efforts as much as engineering efforts can be brought to bear on biology. Thus, the engineering of allosteric ribozymes as molecular sensors and as genetic regulatory switches has served to advance our views of achievable RNA functions and has enabled various technological applications for exploring biologically and biomedically relevant problems.

References

1. Forster AC, Symons RH. Self-cleavage of plus and minus RNAs of a virusoid and a structural model for the active sites. Cell 1987; 49:211-220.
2. Uhlenbeck OC. A small catalytic oligoribonucleotide. Nature 1987; 328:596-600.
3. Hampel A, Tritz R. RNA catalytic properties of the minimum (-)sTRSV sequence. Biochemistry 1989; 28:4929-4933.
4. Sharmeen L, Kuo MTP, Dinter-Gottlieb G et al. Antigenomic RNA of human hepatitis delta virus can undergo self-cleavage. J Virol 1988; 62:2674-2679.
5. Saville BJ, Collins RA. A site-specific self-cleavage reaction performed by a novel RNA in Neurospora mitochondria. Cell 1990; 61:685-96.
6. Winkler WC, Nahvi A, Roth A et al. Control of gene expression by a natural metabolite-responsive ribozyme. Nature 2004; 428:281-286.
7. Teixeira A, Tahiri-Alaoui A, West S et al. Autocatalytic RNA cleavage in the human beta-globin pre-mRNA promotes transcription termination. Nature 2004; 432:526-530.

8. Guerrier-Takada C, Gardiner K, Marsh T et al. The RNA moiety of ribonuclease P is the catalytic subunit of the enzyme. Cell 1983; 35:849-857.

9. Kruger K, Grabowski PJ, Zaug AJ et al. Self-splicing RNA: Autoexcision and autocyclization of the ribosomal RNA intervening sequence of Tetrahymena. Cell 1982; 31:147-157.

10. Peebles CL, Perlman PS, Mecklenburg KL et al. A self-splicing RNA excises an intron lariat. Cell 1986; 44:213-223.

11. Van der Veen R, Arnberg AC, Van der Horst G et al. Excised group II introns in yeast mitochondria are lariats and can be formed by self-splicing in vitro. Cell 1986; 44:225-234.

12. Yean SL, Wuenschell G, Termini J et al. Metal-ion coordination by U6 small nuclear RNA contributes to catalysis in the spliceosome. Nature 2000; 408:881-884.

13. Valadkhan S, Manley JL. Splicing-related catalysis by protein-free snRNAs. Nature 2001; 413:701-707.

14. Noller HF, Hoffarth V, Zimniak L. Unusual resistance of peptidyl transferase to protein extraction procedures. Science 1992; 256:1416-1419.

15. Nissen P, Hansen J, Ban N et al. The structural basis of ribosome activity in peptide bond synthesis. Science 2000; 289;920-930.

16. Moore PB, Steitz TA. The structural basis of large ribosomal subunit funtion. Annu Rev Biochem 2003; 72:813-850.

17. Doudna JA, Cech TR. The chemical repertoire of natural ribozymes. Nature 2002; 418:222-228.

18. Wilson DS, Szostak JW. In vitro selection of functional nucleic acids. Annu Rev Biochem 1999; 68:611-647.

19. Jaschke A. Artificial ribozymes and deoxyribozymes. Curr Opin Struct Biol 2001; 11:321-326.

20. Stroynowski I, Yanofsky C. Transcript secondary structures regulate transcription termination at the attenuator of S. marcescens tryptophan operon. Nature 1982; 298:34-38.

21. Lodmell JS, Dahlberg AE. A conformational switch in Escherichia coli 16S ribosomal RNA during decoding of messenger RNA. Science 1997; 277:1262-1267.

22. Du H, Babitzke P. trp RNA-binding attenuation protein-mediated long distance RNA refolding regulates translation of trpE in Bacillus subtilis. J Biol Chem 1998; 273:20494-20503.

23. Spahn CM, Kleft JS, Grassucci RA et al. Hepatitis C virus IRES RNA-induced changes in the conformation of the 40S ribosomal subunit. Science 2001; 291:1959-1962.

24. Grundy FJ, Winkler WC, Henkin TM. tRNA-mediated transcription antitermination in vitro: Codon-anticodon pairing independent of the ribosome. Proc Natl Acad Sci USA 2002; 99:11121-11126.

25. Valle M, Zavialov A, Li W et al. Incorporation of aminoacyl-tRNA into the ribosome as seen by cryo-electron microscopy. Nat Struct Biol 2003; 10:899-906.

26. Schilling O, Langbein I, Muller M et al. A protein-dependent riboswitch controlling ptsGHI operon expression in Bacillus subtilis: RNA structure rather than sequence provides interaction specificity. Nucleic Acids Res 2004; 32:2853-2864.

27. Gold L, Polinsky B, Uhlenbeck O et al. Diversity of oligonucleotide functions. Annu Rev Biochem 1995; 64:763-797.

28. Osborne SE, Ellington AD. Nucleic acid selection and the challenge of combinatorial chemistry. Chem Rev 1997; 97:349-370.

29. Famulok, M. Oligonucleotide aptamers that recognize small molecules. Curr Opin Struct Biol 1999; 9:324-329.

30. Hermann T, Patel DJ. Adaptive recognition by nucleic acid aptamers. Science 2000; 287:820-825.

31. Mandal M, Breaker RR. Gene regulation by riboswitches. Nat Rev Mol Cell Biol 2004; 5:451-463.

32. Soukup JK, Soukup GA. Riboswitches exert genetic control through metabolite-induced conformational change. Curr Opin Struct Biol 2004; 14:344-349.

33. Nahvi A, Sudarsan N, Ebert MS et al. Genetic control by metabolite binding mRNA. Chem Biol 2002; 9:1043-1049.

34. Winkler WC, Cohen-Chalamish S, Breaker RR. An mRNA structure that controls gene expression by binding FMN. Proc Natl Acad Sci USA 2002; 99:15908-15913.

35. Mironov AS, Gusarov I, Rafikov R et al. Sensing small molecules by nascent RNA: A mechanism to control transcription in bacteria. Cell 2002; 111:747-756.

36. Winkler W, Nahvi A, Breaker RR. Thiamine derivatives bind messenger RNAs directly to regulate bacterial gene expression. Nature 2002; 419:952-956.
37. McDaniel BAM, Grundy FJ, Artsimovitch I et al. Transcription termination control of the S box system: Direct measurement of S-adenosylmethionine by the leader RNA. Proc Natl Acad Sci USA 2003; 100:3083-3088.
38. Winkler WC, Nahvi A, Sudarsan N et al. An mRNA structure that controls gene expression by binding S-adenosylmethionine. Nat Struct Biol 2003; 10:701-707.
39. Grundy FJ, Lehman SC, Henkin TM. The L box regulon: Lysine sensing by leader RNAs of bacterial lysine biosynthesis genes. Proc Natl Acad Sci USA 2003; 100:12057-12062.
40. Sudarsan N, Wickiser JK, Nakamura S et al. An mRNA structure in bacteria that controls gene expression by binding lysine. Genes Dev 2003; 17:2688-2697.
41. Mandal M, Lee M, Barrick JE et al. A glycine-dependent riboswitch that uses cooperative binding to control gene expression. Science 2004; 306:275-279.
42. Mandal M, Boese B, Barrick JE et al. Riboswitches control fundamental biochemical pathways in Bacillus subtilis and other bacteria. Cell 2003; 113:577-586.
43. Mandal M, Breaker RR. Adenine riboswitches and gene activation by disruption of a transcriptional terminator. Nat Struct Mol Biol 2004; 11:29-35.
44. Sudarsan N, Barrick JE, Breaker RR. Metabolite-binding RNA domains are present in the genes of eukaryotes. RNA 2003; 9:644-647.
45. Kubodera T, Watanabe M, Yoshiuchi K et al. Thiamine-regulated gene expression of Aspergillus oryzae thiA requires splicing of the intron containing a riboswitch-like domain in the 5'-UTR. FEBS Lett 2003; 555:516-520.
46. Soukup GA, Breaker RR. Allosteric nucleic acid catalysts. Curr Opin Struct Biol 2000; 10:318-325.
47. Kuwabara T, Warashina M, Taira K. Allosterically controllable ribozymes with biosensor functions. Curr Opin Chem Biol 2000; 4:669-677.
48. Breaker RR. Engineered allosteric ribozymes as biosensor components. Curr Opin Biotechnol 2002; 13:31-39.
49. Silverman SK. Rube Goldberg goes (ribo)nuclear? Molecular switches and sensors made from RNA. RNA 2003; 9:377-383.
50. Tang J, Breaker RR. Rational design of allosteric ribozymes. Chem Biol 1997; 4:453-459.
51. Tang J, Breaker RR. Mechanism for allosteric inhibition of an ATP-sensitive ribozyme. Nucleic Acids Res 1998; 26:4214-4221.
52. Araki M, Okuno Y, Hara Y et al. Allosteric regulation of a ribozyme activity through ligand-induced conformational change. Nucleic Acids Res 1998; 26:3379-3384.
53. Robertson MP, Ellington AD. In vitro selection of an allosteric ribozyme that transduces analytes to amplicons. Nat Biotechnol 1999; 17:62-66.
54. Soukup GA, Breaker RR. Engineering precision RNA molecular switches. Proc Natl Acad Sci USA 1999; 96:3584-3589.
55. Soukup GA, Breaker RR. Design of allosteric hammerhead ribozymes activated by ligand-induced structure stabilization. Structure 1999; 7:783-791.
56. Robertson MP, Ellington AD. Design and optimization of effector-activated ribozyme ligases. Nucleic Acids Res 2000; 28:1751-1759.
57. Jose AM, Soukup GA, Breaker RR. Cooperative binding of effectors by an allosteric ribozyme. Nucleic Acids Res 2001; 29:1631-1637.
58. Kertsburg A, Soukup GA. A versatile communication module for controlling RNA folding and catalysis. Nucleic Acids Res 2002; 30:4599-4606.
59. Wang DY, Lai BH, Sen D. A general strategy for effector-mediated control of RNA-cleaving ribozymes and DNA enzymes. J Mol Biol 2002; 318:33-43.
60. Levy M, Ellington AD. ATP-dependent allosteric DNA enzymes. Chem Biol 2002; 9:417-426.
61. Thompson KM, Syrett HA, Knudsen SM et al. Group I aptazymes as genetic regulatory switches. BMC Biotechnol 2002; 2:21.
62. Atsumi S, Ikawa Y, Shiraishi H et al. Design and development of a catalytic ribonucleoprotein. EMBO J 2001; 20:5453-5460.
63. Hartig JS, Najifi-Shoushtari SH, Grune I et al. Protein-dependent ribozymes report molecular interactions in real time. Nat Biotechnol 2002; 20:717-722.

64. Wang DY, Sen D. Rationally designed allosteric variants of hammerhead ribozymes responsive to the HIV-1 Tat protein. Comb Chem High Throughput Screen 2002; 5:301-312.
65. Vaish NK, Dong F, Andrews L et al. Monitoring post-translational modifications of proteins with allosteric ribozymes. Nat Biotechnol 2002; 20:810-815.
66. Porta H, Lizardi PM. An allosteric hammerhead ribozyme. Biotechnology 1995; 13:161-164.
67. Kuwabara T, Warashina M, Tanabe T et al. A novel allosterically trans-activated ribozyme, the maxizyme, with exceptional specificity in vitro and in vivo. Mol Cell 1998; 2:617-627.
68. Hamada M, Kuwabara T, Warashina M et al. Specificity of novel allosterically trans- and cis-activated connected maxizymes that are designed to suppress BCR-ABL expression. FEBS Lett 1999; 461:77-85.
69. Tanabe T, Takata I, Kuwabara T et al. Maxizymes, novel allosterically controllable ribozymes, can be designed to cleave various substrates. Biomacromolecules 2000; 1:108-117.
70. Komatsu Y, Yamashita S, Kazama N et al. Construction of new ribozymes requiring short regulator oligonucleotides as cofactors. J Mol Biol 2000; 299:1231-1243.
71. Burke DH, Ozerova ND, Nilsen-Hamilton M. Allosteric hammerhead ribozyme TRAPs. Biochemistry 2002; 41:6588-6594.
72. Wang DY, Sen D. A novel mode of regulation of an RNA-cleaving DNAzyme by effectors that bind to both enzyme and substrate. J Mol Biol 2001; 310:723-734.
73. Wang DY, Lai BH, Feldman AR et al. A general approach for the use of oligonucleotide effectors to regulate the catalysis of RNA-cleaving ribozymes and DNAzymes. Nucleic Acids Res 2002; 30:1735-1742.
74. Hartig JS, Grüne I, Najafi-Shoushtari SH et al. Sequence-specific detection of microRNAs by signal-amplifying ribozymes. J Am Chem Soc 2004; 126:722-723.
75. Najafi-Shoushtari SH, Maher G, Famulok M. Sensing complex regulatory networks by conformationally controlled hairpin ribozymes. Nucleic Acids Res 2004; 32:3212-3219.
76. Seetharaman S, Zivarts M, Sudarsan N et al. Immobilized RNA switches for the analysis of complex chemical and biological mixtures. Nat Biotechnol 2001; 19:336-341.
77. Zivarts M, Liu Y, Breaker RR. Engineered allosteric ribozymes that respond to specific divalent metal ions. Nucleic Acids Res 2005; 33:622-631.
78. Koizumi M, Soukup GA, Kerr JNQ et al. Allosteric selection of ribozymes that respond to the second messengers cGMP and cAMP. Nat Struct Biol 1999; 6:1062-1071.
79. Soukup GA, DeRose EC, Koizumi M et al. Generating new ligand-binding RNAs by affinity maturation and disintegration of allosteric ribozymes. RNA 2001; 7:524-536.
80. Srinivasan J, Cload ST, Hamaguchi N et al. ADP-specific sensors enable universal assay of protein kinase activity. Chem Biol 2004; 11:499-508.
81. Soukup GA, Emilsson GAM, Breaker RR. Altering molecular recognition of RNA aptamers by allosteric selection. J Mol Biol 2000; 298:623-632.
82. Piganeau N, Jenne A, Thuillier V et al. An allosteric ribozyme regulated by doxycycline. Angew Chem Int Ed 2000; 39:4369-4373.
83. Piganeau N, Thuillier V, Famulok M. In vitro selection of allosteric ribozymes: Theory and experimental validation. J Mol Biol 2001; 312:1177-1190.
84. Ferguson A, Boomer RM, Kurz M et al. A novel strategy for selection of allosteric ribozymes yields RiboReporter sensors for caffeine and aspartame. Nucleic Acids Res 2004; 32:1756-1766.
85. Robertson MP, Ellington AD. In vitro selection of nucleoprotein enzymes. Nat Biotechnol 2001; 19:650-655.
86. Atsumi S, Ikawa Y, Shiraishi H et al. Selections for constituting new RNA-protein interactions in catalytic RNP. Nucleic Acids Res 2003; 31:661-669.
87. Komatsu Y, Nobuoka K, Karino-Abe N et al. In vitro selection of hairpin ribozymes activated with short oligonucleotides. Biochemistry 2002; 41:9090-9098.
88. Sassanfar M, Szostak JW. An RNA motif that binds ATP. Nature 1993; 364:550-553.
89. Burgstaller P, Famulok M. Isolation of RNA aptamers for biological cofactors by in vitro selection. Angew Chem Int Ed Engl 1994; 33:1084-1087.
90. Jenison RD, Gill SC, Pardi A et al. High-resolution molecular discrimination by RNA. Science 1994; 263:1425-1429.

91. Jiang F, Kumar RA, Jones RA et al. Structural basis of RNA folding and recognition in an AMP-RNA aptamer complex. Nature 1996; 382:183-186.
92. Fan P, Suri AK, Fiala R et al. Molecular recognition in the FMN-RNA aptamer complex. J Mol Biol 1996; 258:480-500.
93. Zimmermann GR, Jenison RD, Wick CL et al. Interlocking structural motifs mediate molecular discrimination by a theophylline-binding RNA. Nat Struct Biol 1997; 4:644-649.
94. Uhlenbeck OC. A small catalytic oligoribonucleotide. Nature 1997; 328:596-600.
95. Haseloff J. Gerlach WL. Simple RNA enzymes with new and highly specific endoribonuclease activity. Nature 1998; 334:585-591.
96. Ruffner DE, Stormo GD, Uhlenbeck OC. Sequence requirements of the hammerhead RNA self-cleavage reaction. Biochemistry 1990; 29:10695-10702.
97. Fedor MJ, Uhlenbeck OC. Kinetics of intermolecular cleavage by hammerhead ribozymes. Biochemistry 1992; 31:12042-12054.
98. Tuschl T, Eckstein F. Hammerhead ribozymes: Importance of stem-loop II for activity. Proc Natl Acad Sci USA 1993; 90:6991-6994.
99. Pley HW, Flaherty K, McKay DM. Three-dimensional structure of a hammerhead ribozyme. Nature 1994; 372:68-74.
100. Scott WG, Finch JT, Klug A. The crystal structure of an all-RNA hammerhead ribozyme: A proposed mechanism for RNA catalytic cleavage. Cell 1995; 81:991-1002.
101. Birikh KR, Heaton PA, Eckstein F. The structure, function and application of the hammerhead ribozyme. Eur J Biochem 1997; 245:1-16.
102. Kurganov BI. Allosteric Enzymes. New York: John Wiley and Sons Ltd., 1978.
103. Perutz M. Mechanisms of cooperativity and allosteric regulation in proteins. New York: Cambridge University Press, 1994.
104. Tang J, Breaker RR. Examination of the catalytic fitness of the hammerhead ribozyme by in vitro selection. RNA 1997; 3:914-925.
105. Frauendorf C, Jaschke A. Detection of small organic analytes by fluorescing molecular switches. Bioorg Med Chem 2001; 9:2521-2524.
106. Sekella PT, Rueda D, Walter NG. A biosensor for theophylline based on fluorescence detection of ligand-induced hammerhead ribozyme cleavage. RNA 2002; 8:1242-1252.
107. Hesselberth JR, Robertson MP, Knudsen SM et al. Simultaneous detection of diverse analytes with an aptazyme ligase array. Anal Biochem 2003; 312:106-112.
108. Steele D, Kertsburg A, Soukup GA. Engineered catalytic RNA and DNA: New biochemical tools for drug discovery and design. Am J Pharmacogenomics 2003; 3:131-144.
109. Tanabe T, Kuwabara T, Warashina M et al. Oncogene inactivation in a mouse model. Nature 2000; 406:473-474.
110. Kuwabara T, Tanabe T, Warashina M et al. Allosterically controllable maxizyme-mediated suppression of progression of leukemia in mice. Biomacromolecules 2001; 2:1220-1228.
111. Soda Y, Tani K, Bai Y et al. A novel maxizyme vector targeting a bcr-abl fusion gene induced cell death in Philadelphia chromosome-positive acute lymphoblastic leukemia. Blood 2004; 104:356-363.
112. Cech TR. Self-splicing of group I introns. Annu Rev Biochem 1990; 59:543-568.
113. Cannone JJ, Subramanian S, Schnare MN et al. The Comparative RNA Web (CRW) Site: An online database of comparative sequence and structure information for ribosomal, intron, and other RNAs. BMC Bioinformatics 2002; 3:2.
114. Werstuck G, Green MR. Controlling gene expression in living cells through small molecule-RNA interactions. Science 1998; 282:296-298.
115. Buskirk AR, Landrigan A, Liu DR. Engineering a ligand-dependent RNA transcriptional activator. Chem Biol 2004; 11:1157-1163.
116. Watanabe T, Sullenger BA. RNA repair: A novel approach to gene therapy. Adv Drug Deliv Rev 2000; 44:109-118.

CHAPTER 2

Ribozymes and Deoxyribozymes Switched by Oligonucleotides

Dipankar Sen* and Edward K.Y. Leung

Abstract

This chapter explores the diverse strategies reported in recent years for using oligonucleotides to switch the catalytic activity of allosteric ribozymes and deoxyribozymes. The earliest allosteric ribozymes were rationally designed and in vitro selected to respond to a variety of small-molecule effectors. However, the possibility of using proteins or RNA or DNA sequences to control the activity of catalytic nucleic acids has always been particularly tantalizing given the potential biomedical utility of such reagents, particularly within living systems. This article reviews the remarkable variety of approaches that have been used to generate ribozymes and deoxyribozymes that can be switched on (and, occasionally, off) by specific RNA or DNA oligonucleotides.

Introduction

The discovery of catalytic RNAs (ribozymes; reviewed in ref. 1) was a milestone in biochemistry in the 1980s. Group I and II self-splicing ribozymes, RNase P, and the four self-cleaving ribozymes—the hammerhcad, hairpin, Varkud satellite, and hepatitis delta virus motifs—have since been investigated exhaustively in terms of both structure and mechanism.[1] Interest in the catalytic possibilities of RNA and of nucleic acids in general, together with advances in the technology of in vitro selection, has resulted also in a variety of artificial catalytic DNAs (deoxyribozymes, DNA enzymes, or DNAzymes), several of which catalyze sequence-dependent RNA cleavage with activities comparable to those of the naturally occurring self-cleaving ribozymes.[2-4]

Protein enzymes often employ the phenomenon of allostery for feedback-mediated control of catalysis (reviewed in ref. 5). However, the naturally occurring ribozymes were found not to use this strategy for control of their activity. To determine whether or not a ribozyme could be allosteric, Tang and Breaker reported in 1997 the rational design and construction of ATP-responsive allosteric hammerhead ribozymes.[6] These ribozymes were constructed by functionally linking a ligand-binding RNA receptor (aptamer) for ATP with a hammerhead ribozyme via a shared helical arm. Several excellent recent reviews on allosteric ribozymes and deoxyribozymes have been published in recent years.[7-12] In a triumphant recapitulation of human-designed allosteric ribozymes, a naturally occurring allosteric ribozyme—the *glmS* ribozyme[13]—has been found as a riboswitch in bacteria (see Chapter 6 on riboswitches).

*Corresponding Author: Dipankar Sen—Department of Molecular Biology and Biochemistry, Simon Fraser University, Burnaby, British Columbia V5A 1S6, Canada. Email: sen@sfu.ca

Nucleic Acid Switches and Sensors, edited by Scott K. Silverman. ©2006 Landes Bioscience and Springer Science+Business Media.

This chapter focuses on the use of RNA/DNA oligonucleotides as allosteric effectors for the switching of ribozyme/deoxyribozyme catalysis. We have categorized the various approaches used by investigators in thematically related paradigms. Although we have tried to emphasize precedence and chronology, this review does not aspire to give a purely chronological account of developments in this field.

Disruptive Inhibition of Natural Ribozymes Using Antisense Oligonucleotides

A number of investigators have explored the use of oligonucleotides as purely antisense effectors for the inhibition of the larger ribozymes such as RNase P and the group I intron ribozyme. Both Childs et al[14] and Grugelsiepe et al[15] have reported that significant and highly specific inhibition of RNase P is possible by the hybridization of short oligonucleotides to the P15 loop (J15/16 bulge loop) within the *E. coli* RNase P RNA. Childs et al[16] and Disney et al[17] reported similar efforts to control the activity of group I ribozymes using antisense oligonucleotide effectors.

Destabilized Ribozymes and Maxizymes

Two approaches conceptually resemble Tang and Breaker's original design of an ATP-dependent hammerhead ribozyme, except using oligonucleotides as effectors. First, Ohtsuka, Komatsu, and coworkers[18-19] severely destabilized the structurally crucial stem II of the hammerhead ribozyme, reducing it to a single G–C base pair and an unstructured loop (Fig. 1A). This was the "default" or inactive state of the ribozyme. Upon addition of the

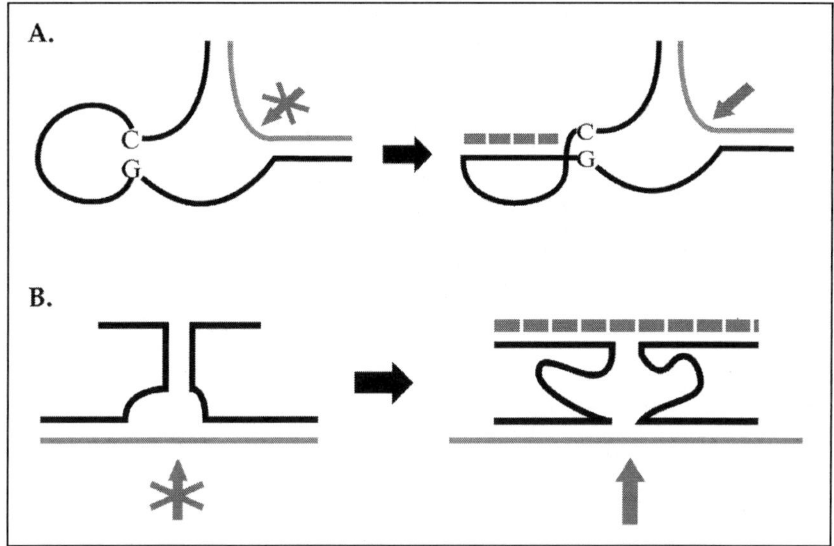

Figure 1. Destabilized ribozymes and Maxizymes. A) Design of an allosteric hammerhead ribozyme by Ohtsuka and coworkers.[18,19] Stem II of the hammerhead ribozyme, which is critical for catalytic activity, has been reduced to a single G–C base pair and a loop. Addition of the oligonucleotide effector (dashed line) contributes to a lengthening and stabilization of stem II via the formation of a pseudo-half-knot. B) An allosteric "Maxizyme" designed by Taira and coworkers.[29] A hammerhead ribozyme motif is created with additional sequence elements that destabilize the catalytic core. Addition of the oligonucleotide effector (dashed line), which hybridizes to portions of this additional sequence, reestablishes a competent catalytic core.

short oligonucleotide effector (shown as a dashed line in Fig. 1A), which was complementary to a portion of the unstructured loop, a pseudo-half-knot structure was formed that helped to increase the effective length and stability of stem II. The unstructured loop here played the role of "aptamer" for the oligonucleotide effector.

Second, Taira and coworkers have published extensively on the design and study of what they have termed "Maxizymes", which are dimeric ribozymes derived from the hammerhead ribozyme.[12,20-34] In the original, elegant Maxizyme design, the catalytic core of the hammerhead was duplicated in such a way that the resultant ribozyme misfolded and was catalytically inactive as a monomer, but folded correctly and was active as a dimer. The active dimeric form, which possesses two sets of substrate-binding arms, was able to cleave at two distinct target RNA sequences. In the evolution of the Maxizyme concept, Taira and coworkers converted one of the two sets of substrate-binding arms to "sensing" arms that could search for aberrant messenger RNA sequences such as from *bcr-abl* fusions.[21] On the basis of such sensing, the Maxizyme would cleave elsewhere within the aberrant mRNA. Eventually the Maxizyme design was modified such that only one competent catalytic core remained, and in the absence of an aberrant mRNA this catalytic core misfolded (Fig. 1B). However, in the presence of the aberrant mRNA that served as the oligonucleotide effector, the catalytic core folded correctly and was able to cleave its target RNA sequence. Thus by convergence, the allosteric form of the Maxizyme shares conceptual similarity with the original allosteric ribozyme of Tang and Breaker.[6]

Of TRAPs and Related Strategies

The goal of the approaches described in this section has been to disable a ribozyme in its "default" state by disruption of its catalytic core via the binding of a *cis*-"attenuator" sequence element to the catalytic core. The addition of a rescuing oligonucleotide effector then disrupts the attenuator-catalytic core interaction by sequestering the attenuator element, thus unmasking the catalytic function.

Porta and Lizardi designed a hammerhead ribozyme construct that has a sequence appended to its 5' substrate-binding arm.[35] The 15-nt terminal portion of this additional sequence may be regarded as an attenuator, because it is complementary to the 5' substrate-binding arm as well as to a significant portion of the ribozyme catalytic core. In the absence of an external 20-nt oligonucleotide effector, the ribozyme folds to a catalytically inactive form due to the complementarity of the attenuator to elements of the ribozyme itself (Fig. 2A). When added, the oligonucleotide effector is able to sequester the attenuator element away from its interaction with the ribozyme core, thus freeing the core to fold properly and also liberating the substrate-binding arms for interacting with the substrate.

By distinguishing an attenuator sequence from an adjacent antisense sequence, Burke and coworkers have made an interesting innovation that they term the TRAP (targeted ribozyme-attenuated probe) method.[36-38] In this approach, the attenuator binds to the ribozyme core, while the antisense sequence permits the initial binding of the effector. In their scheme for catalytic rescue (Fig. 2B), the oligonucleotide effector binds primarily to the antisense part of the ribozyme construct and not necessarily to the attenuator. The chief advantage of this separation of function is an added design versatility for such allosteric ribozymes. By functionally separating the attenuator (which is complementary to the ribozyme core) and the antisense region (which is an arbitrary sequence), the TRAP approach permits the use of any sequence as the oligonucleotide effector.

In analyzing the kinetic properties of these allosteric ribozymes, it was found that substrate binding was not different in the repressed (default) and activated states, consistent with the observation that K_M values did not change upon activation, whereas k_{cat} values did change. Complete rescue of the repressed state was not achieved, because rescue was determined

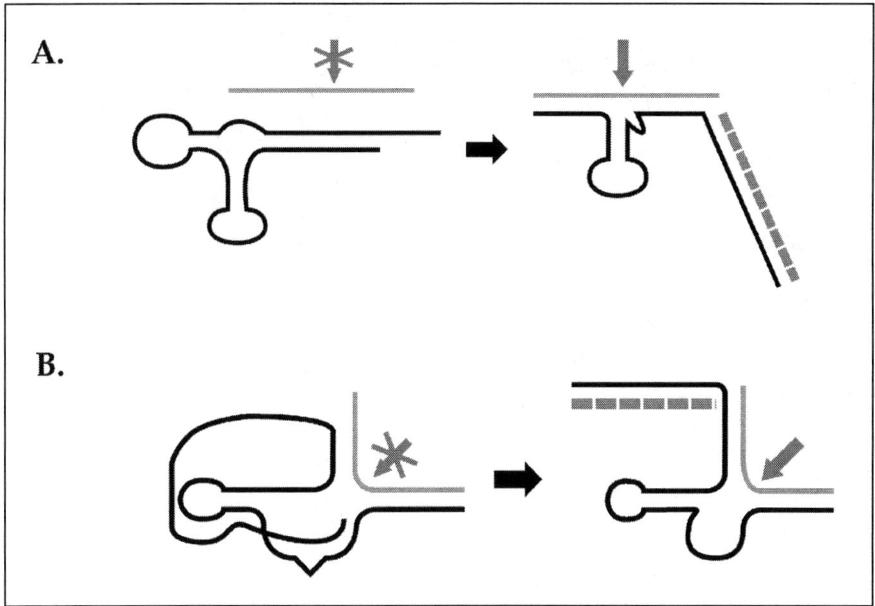

Figure 2. TRAPs and related strategies. A) A version of the hammerhead ribozyme that incorporates an additional 5' sequence extension. In the default state of the ribozyme, this extension sequesters elements of both the ribozyme core and the substrate binding arms.[35] Addition of the oligonucleotide effector (dashed line) restores the catalytically competent core for the ribozyme and frees the arms for binding the substrate. B) The TRAP methodology of Burke and coworkers.[36-38] In the default state, a sequence extension on the 3'-end of a hammerhead ribozyme (i.e., an "attenuator" element) hybridizes with the ribozyme's catalytic core and inactivates it. Addition of the oligonucleotide effector (dashed line) relieves this repression and restores the catalytic activity of the ribozyme. Substrate binding is not substantially impacted in either the repressed or activated states of the ribozyme. Figure continued on next page.

by the relative thermal stabilities of two competing double helices: the attenuator-catalytic core helix ("attenuating helix") and the sense-antisense or sensor-activator helix ("activation helix").

Najafi Shoustari et al[39] reported two distinct constructs based on the hairpin ribozyme, both of which were responsive to a common oligonucleotide effector (Fig. 2C). The difference between the two constructs lies in the fact that rHP-TRAP is catalytically active in the absence of the oligonucleotide effector and can be repressed with the effector, whereas iHP-TRAP is inducible by the oligonucleotide effector. The design strategy for the repressible construct, rHP-TRAP, follows the well-developed antisense idea (described above) that has been used to repress the activity of diverse ribozymes, including the group I intron ribozyme and RNase P.[14-17] Here, an oligonucleotide complementary to a specific unpaired region within the active fold of the ribozyme—in this instance, a loop appended to the hinge region of the hairpin ribozyme—disrupts the catalytically competent structure. In contrast, the design of the inducible iHP-TRAP shares some similarity with the TRAP methodology described by Burke and colleagues,[36-38] because in the repressed state an attenuator element within loop C base-pairs with the substrate binding element within domain A of the ribozyme. Upon addition of an oligonucleotide effector, repression is relieved via the formation of a pseudo-half-knot within loop C, reminiscent of the induction strategy used by Ohtsuka, Komatsu, and colleagues.[18,19]

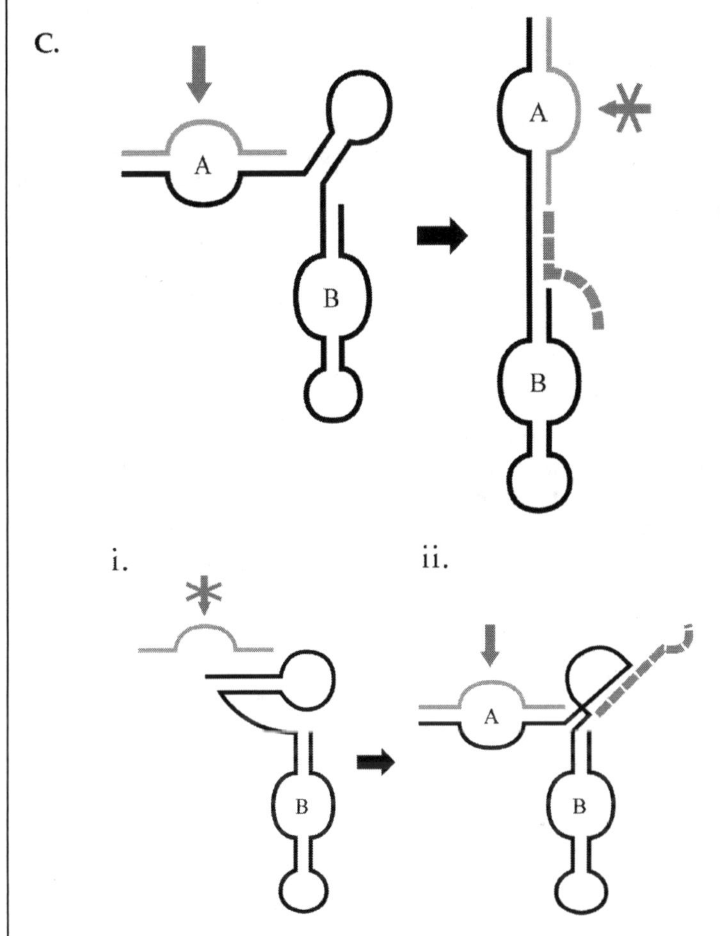

Figure 2, continued. C) Distinct repressible (i) and activatable (ii) constructs of the hairpin ribozyme, as designed by Famulok and coworkers.[39] These ribozymes can be repressed or activated by the same oligonucleotide effector (dashed line).

Interference with Substrate Binding: Expansive Regulation and Deoxyribozyme Logic Gates

A different approach from the TRAP and related methodologies is the use of oligonucleotide effectors not to relieve the disabled catalytic cores of ribozymes but to enable substrate binding. In its default state, such a ribozyme or deoxyribozyme binds substrate poorly or not at all, owing to substrate binding arms that are either of insufficient length or are blocked by competing hybridized sequence elements. Addition of the effector serves to provide additional "effective length" to the binding arms to achieve better substrate binding or else to free the binding arms to enable substrate binding. A major feature of these approaches for ribozyme regulation is that nothing needs to be known about the secondary or tertiary folding of the ribozyme core; instead, control of function is exercised at the level of substrate binding.

Sen and coworkers have reported such an approach (Fig. 3A). Wang and Sen reported a modified design of a 10-23 deoxyribozyme, which instead of the standard 8- or 9-nt substrate binding arms (8 + 8 or 9 + 9 nt in the two arms) had 6 nt in one arm and just 3 nt in the other (6 + 3).[40] Under the RNA cleavage reaction conditions (10 mM Mg^{2+} at 25°C), such a construct was ineffective at binding substrate. However, addition of the oligonucleotide cofactor (Fig. 3A), which binds to both the deoxyribozyme and to the substrate by forming a three-way RNA-DNA junction, supplies a longer substrate binding arm (6 + 7) capable of binding the substrate with much higher affinity. This approach was extended to a variety of ribozymes and deoxyribozymes, including the hammerhead ribozyme, the 8-17 deoxyribozyme, and the bipartite DNAzyme.[41] This overall approach at effector-mediated control of nucleic acid enzymes has been termed "expansive regulation" to distinguish it from classic allosteric regulation, which typically impacts the chemical step (rather than the substrate-binding step) of the enzymatic reaction. In a modification of this approach (Fig. 3B), an ATP-binding aptamer was incorporated into modified ribozymes to make them responsive to ATP or adenosine as an effector.[42] Recently, Sando et al have adapted this approach to generate an mRNA "sensing" system.[43]

Stojanovic and coworkers[44-46] have been making molecular switches from oligonucleotide-controlled allosteric deoxyribozymes, and they have recently reported highly innovative designs for the construction of deoxyribozyme-based logic gates. The designs of these allosteric ribozymes involve oligonucleotide effectors that serve as either inducers or repressors, sometimes within the same deoxyribozyme construct. Figure 3C shows the design of the $i_1AND i_3ANDNOTi_6$ deoxyribozyme, which shows the "logic" behaviour of being turned on in the presence of *both* effectors i and ii (but not in the presence of only one) and being turned off by effector iii, regardless of the presence of effectors i and ii. The activation portions of the design of this deoxyribozyme echo the TRAP methodology, except that here the attenuator sequence elements (defined by the i_1 loop and its 5' continuation as well as the i_3 loop and its 3' continuation) base-pair with and inactivate the substrate binding arms (rather than the catalytic core) of the deoxyribozyme, with the i_1 and i_3 loops acting as sensors. Addition of oligonucleotide effectors complementary to i_1 and i_3 leads to freeing of the deoxyribozyme substrate binding arms and an increase in catalytic activity. To enable oligonucleotide-mediated repression of the same deoxyribozyme construct, the authors engineered a stem-loop structure, i_6, branching from the catalytic core. This stem-loop structure is carefully positioned such that its presence does not perturb the default activity of the deoxyribozyme. Repression is achieved by addition of an oligonucleotide effector that hybridizes to i_6 and disrupts the catalytically relevant fold of the deoxyribozyme core.

Defective Ribozymes and "Half"-Ribozymes

These strategies for generating allosteric ribozymes use an incomplete or defective ribozyme as their default states. The oligonucleotide effector then supplies a complementation function, by either rectifying one or more key mutations in the catalytic core of the ribozyme or by supplying a missing portion of the ribozyme.

Vauleon and Müller created a defective default version (Fig. 4A, upper) of the hairpin ribozyme. In their construct HP-G25, a key C-to-G point mutation that abolished all catalytic activity was incorporated into the catalytic core within bulge-loop B.[47] The addition of an oligonucleotide effector (EF-01) resulted in the displacement of a stretch of the ribozyme that contained the deleterious mutation. The effector, which incorporated the C nucleotide necessary for catalytic activity, was able to complement the defective ribozyme and restore the activity by regenerating a catalytically functional core capable of docking with the substrate bulge loop A. Unexpectedly, when the EF-01 effector was added back to a more struc-

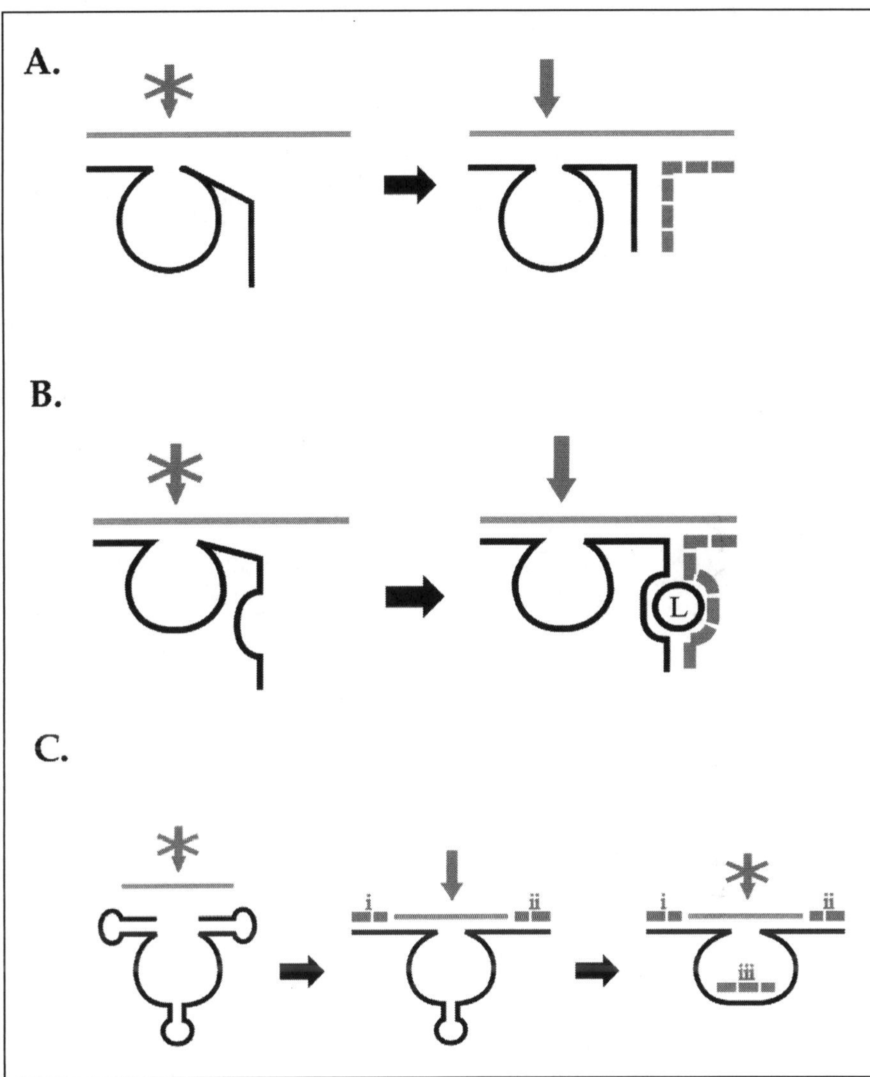

Figure 3. Expansive control and logic gates. A) The "expansive" approach of Sen and coworkers.[40-42] In the default state, the substrate-binding capability of a ribozyme or deoxyribozyme is compromised by the presence of a substrate binding arm of inadequate length. Addition of the oligonucleotide effector (dashed line), which binds both enzyme and substrate, enlarges the effective length of the enzyme's substrate binding arms and thus enhances catalysis. B) The three-helix junction design of expansively controlled ribozymes can be adapted, via inclusion of a ligand-binding aptamer, to make the ribozyme responsive to both an oligonucleotide and a small-molecule effector. C) Deoxyribozyme logic gates as designed by Stojanovic and coworkers.[44-46] In its default state, the enzyme is repressed by foldback sequence elements that sequester the two substrate-binding arms. This repression can be relieved by the addition of oligonucleotide effectors i and ii (dashed lines). However, addition of a further effector, iii, that is complementary to a stem-loop built into the enzyme's catalytic core destroys the catalytically competent fold of the catalytic core, despite the presence of the activating effectors i and ii.

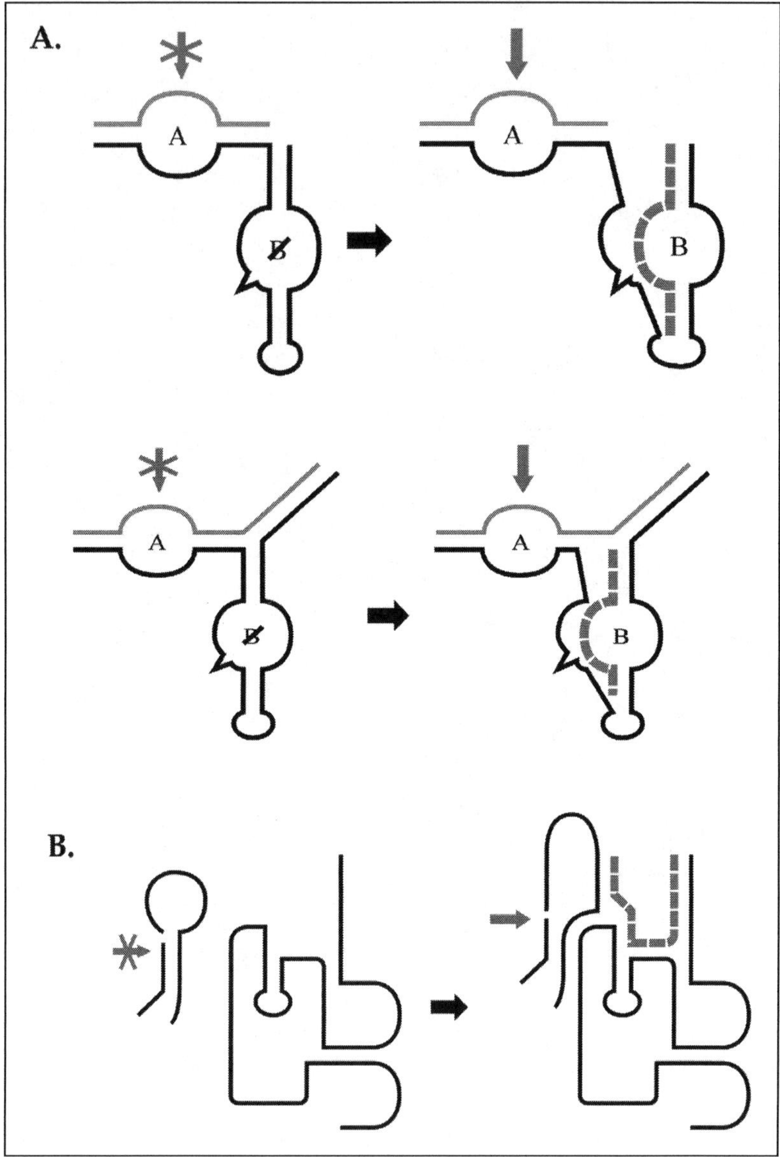

Figure 4. Defective ribozymes and half-ribozymes. A) Two defective versions of the hairpin ribozyme created by Vauleon and Müller.[47] In their default states, these ribozymes contain a key point mutation in bulge-loop B that abolishes catalytic activity. Addition of the oligonucleotide effector (dashed line) in each case corrects the defect and reestablishes catalytic activity. B) Design by Seiwert and coworkers of a half-ribozyme[48,49] based on the in vitro-selected class I RNA ligase.[51] In its default state, significant portions of the catalytic core and substrate-binding arm of the ribozyme are missing, rendering it catalytically inactive. Complementation by an oligonucleotide effector (dashed line), which supplies the missing portions of the catalytic core as well as substrate-binding arm, restores the catalytic activity of the ribozyme.

turally constrained version of the defective hairpin ribozyme, the TW-G25 construct (Fig. 4A, lower), the effector was still able to complement and rescue catalysis.

Seiwert and colleagues[48,49] have been working with the in vitro-selected class I RNA ligase, which was first reported by Bartel and coworkers.[51] The class I ligase has a secondary structure consisting of seven clearly defined helical regions separated by nonhelical elements; of these, the P1 stem lies wholly within the "substrate" RNA. The pair of substrate RNAs to be ligated are in turn bound to the ribozyme at the P2 stem. The class I ligase was deconstructed in such a way that the default enzyme contained only the complete stems P3, P5, and P6 along with sequence elements that compose the P4 and P7 stems within the complete ribozyme (Fig. 4B). As expected, the severely truncated default ribozyme had essentially no detectable ligase activity when the substrate RNAs were added. However, complementation with the RNA oligonucleotide effector structurally completed the P4 and P7 stems and supplied the substrate-docking site at the P2 stem, thereby restoring activity.

The elegance and utility of the half-ribozyme approach is that the oligonucleotide effector, whose sequence can be modified and evolved using in vitro selection methods, can in principle be an important RNA sequence, such as a 5'-untranslated region (5'-UTR) from the hepatitis C virus (HCV). The class I ligase half-enzyme, when complemented by the HCV RNA effector molecule, serves as a highly sensitive indicator for the presence of the HCV RNA. For example, $\sim 10^9$-fold enhancement of catalytic rate by a class I ligase in the presence of the HCV RNA effector was reported.[48] Using this approach, zeptomole quantities (10^{-21} mol, i.e., ~ 1000 molecules) of HCV RNA could be detected, as monitored by RNA cleavage activity. This group's second paper reported that three commonly found sequence variants of the HCV 5'-UTR were comparably efficient at reconstituting the ligase ribozyme.[49] When the reconstituted ribozyme activity was coupled with two different detection methods (an ELISA-based method and an optical immunoassay), both permitted detection of as few as $\sim 10^6$ HCV RNA molecules.

Unique Systems

Figure 5A shows an in vitro selected allosteric ribozyme that is responsive to oligonucleotide effectors. In 1999, Robertson and Ellington reported the selection of a new RNA ligase named L1.[50] The catalytic activity of the L1 ligase was $\sim 10,000$-fold higher in the presence of a cDNA primer that was present during the in vitro selection process. As initially isolated, the L1 ligase was dependent on this primer as an oligonucleotide effector. Following truncation and replacement of nonconserved residues by an adenosine-specific aptamer, the new construct L1.dB2-ATPm1 (Fig. 5B) was now adenosine/ATP-responsive in the presence of the bound cDNA primer. For example, catalytic activity was stimulated ~ 800-fold by 1 mM ATP.

Perspective

In this chapter we have summarized the variety of creative ways in which investigators have used oligonucleotides to influence the catalytic activity of both natural and in vitro selected ribozymes and deoxyribozymes. Undoubtedly, the strategies summarized here are not the only ones that will succeed, and we expect that others will be reported in the years to come. The potential utility is promising for using these controllable nucleic acid enzymes as both sensors and in vivo therapeutic agents. In the latter respect, the Maxizymes developed by Taira and coworkers[12,20-34] have already been shown to effective in vivo.[29,32,34] We anticipate that the in vivo effectiveness of many of the others strategies will also be reported soon.

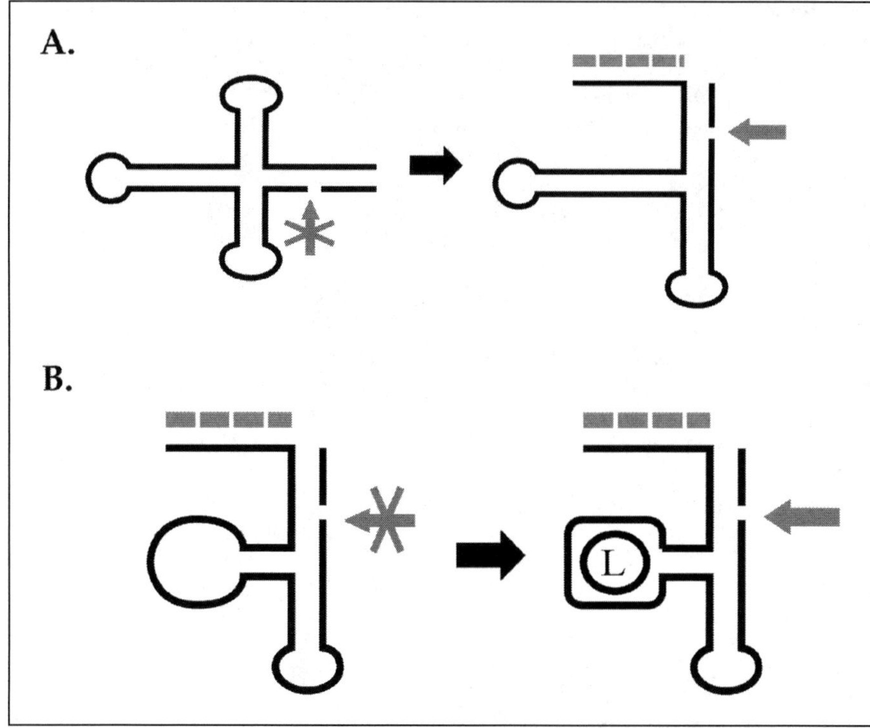

Figure 5. An in vitro-selected system. A) An RNA ligase ribozyme in vitro selected by Robertson and Ellington to be dependent on an oligonucleotide effector.[50] B) A portion of the ribozyme's structure can be replaced by a ligand-binding aptamer, which renders the ribozyme responsive both to oligonucleotides and to small-molecule effectors such as ATP.

References

1. Doudna AJ, Cech TR. The chemical repertoire of natural ribozymes. Nature 2002; 418:222-228.
2. Breaker RR, Joyce GF. A DNA enzyme that cleaves RNA. Chem Biol 1994; 1:223-229.
3. Santoro SW, Joyce GF. A general purpose RNA-cleaving DNA enzyme. Proc Natl Acad Sci USA 1997; 94:4262-4266.
4. Feldman AR, Sen D. A new and efficient DNA enzyme for the sequence-specific cleavage of RNA. J Mol Biol 2001; 313:283-294.
5. Perutz M. Mechanisms of cooperativity and allosteric regulation in proteins. New York: Cambridge University Press, 1994.
6. Tang J, Breaker RR. Rational design of allosteric ribozymes. Chem Biol 1997; 4:453-459.
7. Soukup GA, Breaker RR. Engineering precision RNA molecular switches. Proc Natl Acad Sci USA 1999; 96:3584-3589.
8. Soukup GA, Breaker RR. Design of allosteric hammerhead ribozymes as activated by ligand-induced structure stabilization. Structure Fold Des 1999; 7:783-791.
9. Breaker RR. Engineered allosteric ribozymes as biosensor components. Curr Opin Biotech 2002; 13:31-39.
10. Rajendran M, Ellington AD. Selecting nucleic acids for biosensor applications. Comb Chem High Throughput Screen 2002; 5:263-270.
11. Silverman SK. Rube Goldberg goes (ribo)nuclear? Molecular switches and sensors made from RNA. RNA 2003; 9:377-83.

12. Warashina M, Kuwabara T, Taira K. Working at the cutting edge: The creation of allosteric ribozymes. Structure Fold Des 2000; 8:R207-212.

13. Winkler WC, Nahvi A, Roth A et al. Control of gene expression by a natural metabolite-responsive ribozyme. Nature 2004; 428:281-286.

14. Childs JL, Poole AW, Turner DH. Inhibition of Escherichia coli RNase P by oligonucleotide directed misfolding of RNA. RNA 2003; 9:1437-1445.

15. Gruegelsiepe H, Willkomm DK, Goudinakis O et al. Antisense inhibition of Escherichia coli RNase P RNA: Mechanistic aspects. ChemBioChem 2003; 4:1049-1056.

16. Childs JL, Disney MD, Turner DH. Oligonucleotide directed misfolding of RNA inhibits Candida albicans group I intron splicing. Proc Natl Acad Sci USA 2002; 99:11091-11096.

17. Disney MD, Childs JL, Turner DH. New approaches to targeting RNA with oligonucleotides: Inhibition of group I intron self-splicing. Biopolymers 2004; 73:151-161.

18. Komatsu Y, Yamashita S, Kazama N et al. Construction of new ribozymes requiring short regulator oligonucleotides as a cofactor. J Mol Biol 2000; 299:1231-1243.

19. Komatsu Y, Ohtsuka E. Regulation of ribozyme cleavage activity by oligonucleotides. Methods Mol Biol 2004; 252:165-177.

20. Kuwabara T, Warashina M, Tanabe T et al. A novel allosterically trans-activated ribozyme, the maxizyme, with exceptional specificity in vitro and in vivo. Mol Cell 1998; 2:617-627.

21. Hamada M, Kuwabara T, Warashina M et al. Specificity of novel allosterically trans- and cis-activated connected maxizymes that are designed to suppress BCR-ABL expression. FEBS Lett 1999; 461:77-85.

22. Nakayama A, Kuwabara T, Warashina M et al. CTAB-mediated enrichment for active forms of novel dimeric maxizymes. FEBS Lett 1999; 448:67-74.

23. Kuwabara T, Warashina M, Nakayama A et al. tRNA^{Val}-heterodimeric maxizymes with high potential as geneinactivating agents: Simultaneous cleavage at two sites in HIV-1 Tat mRNA in cultured cells. Proc Natl Acad Sci USA 1999; 96:1886-91.

24. Tanabe T, Takata I, Kuwabara T et al. Maxizymes, novel allosterically controllable ribozymes, can be designed to cleave various substrates. Biomacromolecules 2000; 1:108-17.

25. Kuwabara T, Warashina M, Taira K. Allosterically controllable maxizymes cleave mRNA with high efficiency and specificity. Trends Biotechnol 2000; 18:462-468.

26. Zhou JM, Nakamatsu Y, Kuwabara T et al. Chemical and enzymatic probing of effector-mediated changes in the conformation of a maxizyme. J Inorg Biochem 2000; 78:261-268.

27. Kuwabara T, Tanabe T, Warashina M et al. Allosterically controllable maxizyme-mediated suppression of progression of leukemia in mice. Biomacromolecules 2001; 2:1220-1228.

28. Kuwabara T, Hamada M, Warashina M et al. Allosterically controlled single-chained maxizymes with extremely high and specific activity. Biomacromolecules 2001; 2:788-99.

29. Iyo M, Kawasaki H, Taira K. Allosterically controllable maxizymes for molecular gene therapy. Curr Opin Mol Ther 2002; 4:154-65.

30. Nakayama A, Warashina M, Kuwabara T et al. Effects of cetyltrimethylammonium bromide on reactions catalyzed by maxizymes, a novel class of metalloenzymes. J Inorg Biochem 2002; 78:69-77.

31. Kuwabara T, Warashina M, Taira K. Cleavage of an inaccessible site by the maxizyme with two independent binding arms: An alternative approach to the recruitment of RNA helicases. J Biochem 2002; 132:149-155.

32. Oshima K, Kawasaki H, Soda Y et al. Maxizymes and small hairpin-type RNAs that are driven by a tRNA promoter specifically cleave a chimeric gene associated with leukemia in vitro and in vivo. Cancer Res 2003; 63:6809-6814.

33. Iyo M, Kawasaki H, Taira K. Maxizyme technology. Methods Mol Biol 2004; 252:257-265.

34. Soda Y, Tani K, Bai Y et al. A novel maxizyme vector targeting a bcr-abl fusion gene induced specific cell death in Philadelphia chromosome-positive acute lymphoblastic leukemia. Blood 2004; 104:356-63.

35. Porta H, Lizardi PM. An allosteric hammerhead ribozyme. Biotechnology 1995; 13:161-164.

36. Burke DH, Ozerova ND, Nilsen-Hamilton M. Allosteric hammerhead ribozyme TRAPs. Biochemistry 2002; 41:6588-6594.

37. Saksmerprome V, Burke DH. Structural flexibility and the thermodynamics of helix exchange constrain attenuation and allosteric activation of hammerhead ribozyme TRAPs. Biochemistry 2003; 42:13879-13886.

38. Saksmerprome V, Burke DH. Deprotonation stimulates productive folding in allosteric TRAP hammerhead ribozymes. J Mol Biol 2004; 341:685-694.

39. Najafi-Shoushtari SH, Mayer G, Famulok M. Sensing complex regulatory networks by conformationally controlled hairpin ribozymes. Nucleic Acids Res 2004; 32:3212-3219.

40. Wang DY, Sen D. Expansive regulation of an RNA-cleaving DNAzyme by effectors that bind to both enzyme and subsrate. J Mol Biol 2001; 310:723-734.

41. Wang DY, Lai B, Sen D. A general approach for the use of oligonucleotide effectors to regulate the catalysis of RNA-cleaving ribozymes and DNAzymes. Nucleic Acids Res 2002; 30:1735-1742.

42. Wang DY, Lai B, Sen D. A general strategy for effector-mediated control of RNA-cleaving ribozymes and DNA enzymes. J Mol Biol 2002; 318:33-43.

43. Sando S, Sasaki T, Kanatani K et al. Amplified nucleic acid sensing using programmed self-cleaving DNAzyme. J Am Chem Soc 2003; 125:15720-15721.

44. Stojanovic MN, Mitchell TE, Stefanovic D. Deoxyribozyme-based logic gates. J Am Chem Soc 2002; 124:3555-3561.

45. Stojanovic MN, Stefanovic D. Deoxyribozyme-based half-adder. J Am Chem Soc 2003; 125:6673-6676.

46. Stojanovic MN, Stefanovic D. A deoxyribozyme-based molecular automaton. Nat Biotechnol 2003; 21:1069-1074.

47. Vauleon S, Muller S. External regulation of hairpin ribozyme activity by an oligonucleotide effector. ChemBioChem 2003; 4:220-224.

48. Vaish NK, Jadhav VR, Kossen K et al. Zeptomole detection of a viral nucleic acid using a target-activated ribozyme. RNA 2003; 9:1058-1072.

49. Kossen K, Vaish NK, Jadhav VR et al. High-throughput ribozyme-based assays for detection of viral nucleic acids. Chem Biol 2004; 11:807-815.

50. Robertson MP, Ellington AD. In vitro selection of an allosteric ribozyme that transduces analytes to amplicons. Nat Biotechnol 1999; 17:62-66.

51. Ekland EH, Szostak JW, Bartel DP. Structurally complex and highly active RNA ligases derived from random RNA sequences. Science 1995; 269:364-370.

CHAPTER 3

Ribozymes Switched by Proteins

Tan Inoue* and Yoshiya Ikawa

Abstract

Specific RNA-binding proteins regulate RNA conformations in protein-switched ribozymes. Ribozyme activity can be regulated with RNA-binding proteins, as in the case of natural RNP ribozymes like RNase P that consist of an RNA subunit and an RNA-binding protein. Recent studies have shown that both natural and artificial catalytic RNAs are convertible to allosterically regulated ribozymes by molecular design at the secondary structure or three-dimensional levels and also by performing selections from combinatorial libraries. In protein-switched ribozymes, the specific binding of a protein stabilizes a particular conformation of RNA in active or inactive form such that ribozyme activity is regulated by the presence or absence of the RNA-binding protein. In applications, a protein-regulated ribozyme can be utilized as a biosensor that detects the concentration of the binding proteins in vitro and in vivo. Further advances in the design and construction of protein-regulated ribozymes are expected because of rapid progress in structural biology. Future studies will also shed new light on the molecular evolution pathways of ribozymes.

Introduction

Most functional RNAs in cells are physically associated with protein molecules to form a ribonucleoprotein (RNP) complex. The structure of RNA in an RNP is stabilized via RNA-protein interactions that promote catalysis and other tasks required for gene expression. RNA itself folds hierarchically starting with double-helix formation based on Watson-Crick base pairs and the G-U wobble base pair.[1] Because a number of base-pairing combinations are available, the same RNA molecule can form many alternative structures. Because these alternative conformations are often equally stable, the stabilization of a particular conformation is needed to give an RNA molecule its characteristic structure and allow it to perform its functional role. A high concentration of cations is particularly effective for stabilizing a unique RNA conformation.[2,3] However, high cation concentrations can also promote unwanted conformations, and therefore this approach is impractical in vivo. To overcome this dilemma, a protein that can specifically bind to a particular RNA conformation can be employed to enable a unique RNA function. In many cases, cationic residues such as arginine and lysine in proteins are involved in intermolecular interactions with negatively charged RNA surfaces.

Nature has produced catalytic RNAs that are regulated by RNA-binding proteins. As a prelude to discussing artificial protein-switched ribozymes, here we describe two well-studied

*Corresponding Author: Tan Inoue—Graduate School of Biostudies, Kyoto University, Kyoto 606-8502, Japan. Email: tan@kuchem.kyoto-u.ac.jp

Nucleic Acid Switches and Sensors, edited by Scott K. Silverman. ©2006 Landes Bioscience and Springer Science+Business Media.

natural RNPs with enzymatic function, RNase P and the ribosome. RNase P catalyzes specific phosphodiester bond hydrolysis in pre-tRNAs to produce mature tRNA 5'-ends.[4-6] The examples of RNase P identified so far are holoenzymes comprising one RNA molecule and at least one protein molecule. A bacterial RNase P typically consists of a large RNA subunit (350-400 nucleotides; 100-130 kDa) and one small protein subunit (~120 amino acids; 12-13 kDa).[7] The active site of the enzyme is contained in RNA components that function in vitro without protein in the presence of a high concentration of cations.[8] However, a protein component is required for catalysis under physiological conditions. It is believed that the protein component stabilizes the holoenzyme against electrostatic distortion and increases the turnover rate by specifically binding the substrate over the product.[2,7,9]

The RNA component of bacterial RNase P consists of two major domains, catalytic (C) and specificity (S).[10-12] RNase P variants can be categorized into two types, A and B, on the basis of sequence characteristics of peripheral elements located outside the core domains of conserved C and S.[12-14] The most rigorously investigated RNase Ps are those from *E. coli* and *B. subtilis* that belong to types A and B, respectively. The X-ray crystal structures of the S domains of types A and B RNase P RNAs from *T. thermophilus* and *B. subtilis*, respectively, have been solved. These structures offer insights into the evolution of the architecture of RNase P RNA as an assemblage of modular structural components.[15,16]

Among RNase P proteins consisting of ~120 amino acids, ~40 amino acids are conserved or nearly so.[17] Consistent with phylogenic data, both the A-type and B-type protein can activate the other type of RNA in vitro. Furthermore, the A-type protein can complement B-type RNA in vivo.[2] Thus the two types of RNA and protein are expected to share structural elements that participate in RNA-protein interactions for the activation of RNA. Structural data on the protein subunits of A and B types are available.[7,17,18] Both A-type and B-type proteins share high structural similarity in that they are a mixed α-β protein with an unusual left-handed β-α-β turn, which is similar to ribosomal protein S9, domain II of ribosomal protein S5, and also domain IV of elongation factor EF-G. In the case of S5 and EF-G, the left-handed β-α-β turn contains an RNP box motif responsible for RNA binding. Since the RNP box motif is the most conserved region among RNase P proteins, these proteins primarily recognize RNase P RNAs via direct interactions with RNA's phylogenically conserved core. In addition, the protein is also known to interact with the pre-tRNA leader region, indicating that this interaction increases the turnover rate.[19-21] Recently the crystal structure of the C-domain of type B RNase P from *B. stearothermophilus* was determined (N. R. Pace, personal communication). This achievement enables us to view not only the whole structure of RNase P ribozyme but also its ternary complex with its protein cofactor and the substrate tRNA.

In bacterial ribosomes, approximately two-thirds and one-third of the mass consist of RNA and protein, respectively.[22] The proteins are named according to the subunit of the ribosome in which they are found, such that they belong to the small (S1 to S31) and large (L1 to L44) subunits. Most of the proteins are located at the core of ribosomal RNA (rRNA). Ribosomal proteins, particularly those of the large subunit, are composed of a globular, surface-exposed domain with long finger-like projections that extend into the rRNA core to stabilize its conformation.[23-25] Generally, the proteins interact with multiple RNA elements in different domains. In the large subunit, approximately one-third of the 23S rRNA nucleotides are in van der Waals contact with protein, and the L22 protein associates with all six domains of the 23S RNA. Proteins termed S4 and S7 promote the assembly of 16S rRNA by interacting with two helices in the RNA.[25] Thus proteins are responsible for the organization and stabilization of the higher-order rRNA structure, thereby finely tuning the structure for optimal function. It has been shown that 16S and 23S rRNA are responsible for decoding and peptide synthesis, respectively,[24,26,27] indicating that proteins play an active role to conduct the entire translation process smoothly.

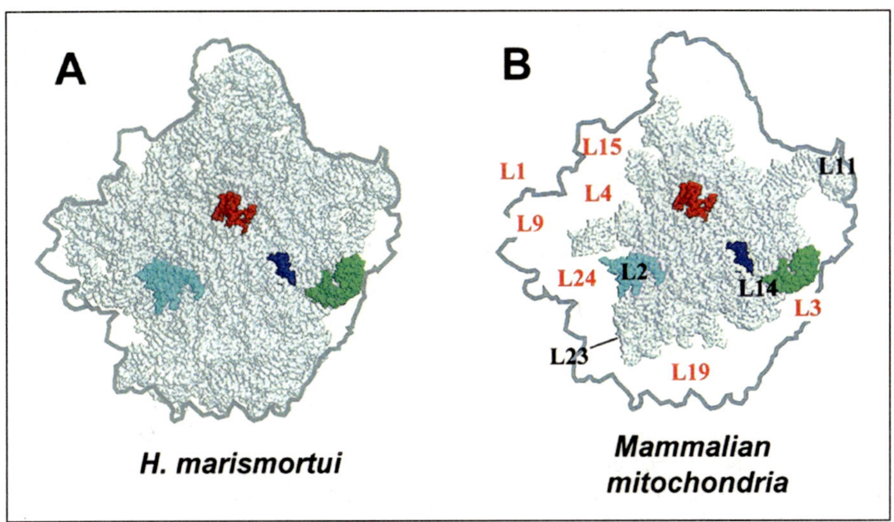

Figure 1. 3D model for a mitochondrial large ribosomal RNA based on the crystal structure of the 50S subunit. All atomic coordinates of the *H. marismortui* 50S subunit were obtained from the Protein Data Bank (pdb id 1FFK). A) A model structure for the mitochondrial ribosome was constructed by removal of nonconserved rRNA portions from the atomic coordinates of 1FFK based on the secondary structure of the large ribosomal RNA from bovine mitochondria. B) The outline shows an edge line of the crystal structure of the 50S subunit from the crown view. Some functional rRNA domains have been colored as follows: red, P loop; blue, A loop; green, S/R loop; light blue, L2 binding helix (H66). The topological orientation of the ribosomal protein is shown on the model for the bovine mitochondrial ribosome. (Reproduction from ref. 28 with permission.)

A comparison of the ribosome from mammalian mitochondria and *E. coli* revealed an interesting evolutionary relationship between rRNA and ribosomal protein. The two ribosomes have approximately the same molecular weight, but the RNA:protein ratios of the mitochondrial and *E. coli* ribosome are 1:2 and 2:1, respectively (Fig. 1). This suggests that during evolution, certain RNA components were replaced by protein components in mitochondria. The structural analyses of these ribosomes clarified that proteins compensated for shortened or lost portions of rRNAs.[28,29]

In this chapter, we will describe artificial protein-switched ribozymes for which the enzymatic function of RNA is under the allosteric control of RNA-binding proteins. Rational molecular design at the atomic level and combinatorial techniques were employed for ribozyme construction. Three classes of protein-switched ribozymes will be presented here: the group I intron ribozyme, the hammerhead ribozyme, and the artificial L1 RNA ligase. The group I intron ribozyme was engineered to mimic the evolution of a ribozyme into an RNP, whereas the hammerhead ribozyme and L1 ligase were engineered for biotechnology applications.

At present, the regulation of protein function is less sophisticated in artificially developed protein-switched ribozymes when compared with their natural counterparts. In the natural systems, highly evolved proteins control the entire catalysis or translation process. However, it has become clear that simple molecular design and/or selection for incorporating naturally occurring RNA-protein interactions into a natural or artificial ribozyme is highly effective for regulating ribozyme function. We expect that the rather primitive artificial proteins of the present will continue to be refined with advanced molecular design and selection techniques that mimic evolutionary pathways.

From a Self-Splicing Group I Intron RNA to a Self-Splicing RNP

Self-splicing group I intron RNAs possess a catalytic core that is responsible for catalyzing RNA splicing reactions. In vivo, most group I introns are thought to be associated with specific proteins that stabilize the active RNA conformation.[30] For example, CYT-18, a tRNA synthetase of *Neurospora crassa*, functions as a group I intron specific splicing factor by binding to the conserved P4-P6-P6a region of various group I introns.[31,32] In the *Tetrahymena* group I intron RNA, its large peripheral domain termed P5abc interacts with the P4-P6-P6a region. This peripheral domain is important for performing efficient splicing reactions under physiological conditions;[33,34] its deletion drastically reduces activity. However, the addition of CYT-18 to a reaction mixture containing a *Tetrahymena* RNA derivative lacking P5abc results in considerable RNA activation due to formation of specific RNA-protein interactions. This indicates that group I intron RNAs share conserved and interchangeable modular units.[32]

On the basis of this observation, the *Tetrahymena* group I intron RNA was converted to a self-splicing RNP by employing a molecular modeling technique.[35] The RNP was designed to fix two newly introduced structural elements in a derivative of the intron RNA that interacts with an artificial protein (Fig. 2A). The conserved P4-P6 domain, which has a hairpin-shaped structure, serves as a scaffold for correct folding of the active form of the intron RNA. In P4-P6, specific interactions between the terminal P5b tetraloop and a tetraloop receptor within P6 act as a clamp to enforce the overall structure (Fig. 2A, left).[36-40] Disruption of the tetraloop-receptor interaction causes a significant loss of splicing activity.[35] Fixation of these interacting elements with the designed protein is expected if the protein can appropriately bind to the RNA. An RNA derivative was devised by replacing the terminal P5b tetraloop and the internal receptor loop in P6, which directly interact in the original RNA, with two peptide-binding motifs, boxB and RRE from bacteriophage λ and HIV-1, respectively (Fig. 2A, right).[41,42] The structure of the RNA-binding peptide was superimposed on the corresponding peptide-binding motif in a computationally designed model RNA derivative. In the modeled structure, two terminal regions consisting of RNA-binding peptides, λN1-19 and HIV-1 Rev34-50 that respectively bind to boxB and RRE, were joined via a linker peptide consisting of four consecutive alanines (Fig. 2A, right). High-resolution structures of RNA-peptide interactions employed in the design have been characterized by NMR spectroscopy.[43,44]

The binding of the designed protein facilitates RNA splicing reactions both in vitro and in vivo. The final yield of the splicing reaction increased up to eight-fold in the presence of the corresponding protein.[35] However, the protein is not involved in the reaction steps, because K_m for the guanosine cofactor and k_{cat} were barely influenced by the presence of the protein.[35] Two natural group I intron splicing factors, CBP2 and CYT-18, are also known to bind to intron RNAs to stabilize their active forms by facilitating RNA folding.[45] Interestingly, these two proteins can increase k_{cat} up to several hundred times compared with RNA alone, as determined with the bi5 intron RNA.[46,47]

In our modeled structure, the two terminal regions of the designed proteins were joined via four consecutive alanine residues, Ala-Ala-Ala-Ala (which form part of a stable α-helix). The protein with Ala-Ala-Ala-Ala linker functions twice as effectively as the structurally flexible Gly-Gly-Gly-Gly linker (eight-fold versus four-fold activation), although both facilitate folding. The optimized linker sequence was Gly-Val-Gly-Arg, which was determined using in vivo selection because modeling is impractical for such optimization.[35] The protein with the selected sequence that binds to RNA with high affinity facilitated RNA folding much better than the original Ala-Ala-Ala-Ala linker in vitro. A different selection was performed to obtain alternative forms of the Rev-RRE interaction.[48] Newly established RNA-protein interactions efficiently regulate and enhance splicing reactions in vivo and in vitro. The affinity between RNA and the selected protein was comparable to the Rev-RRE interaction, demonstrating that

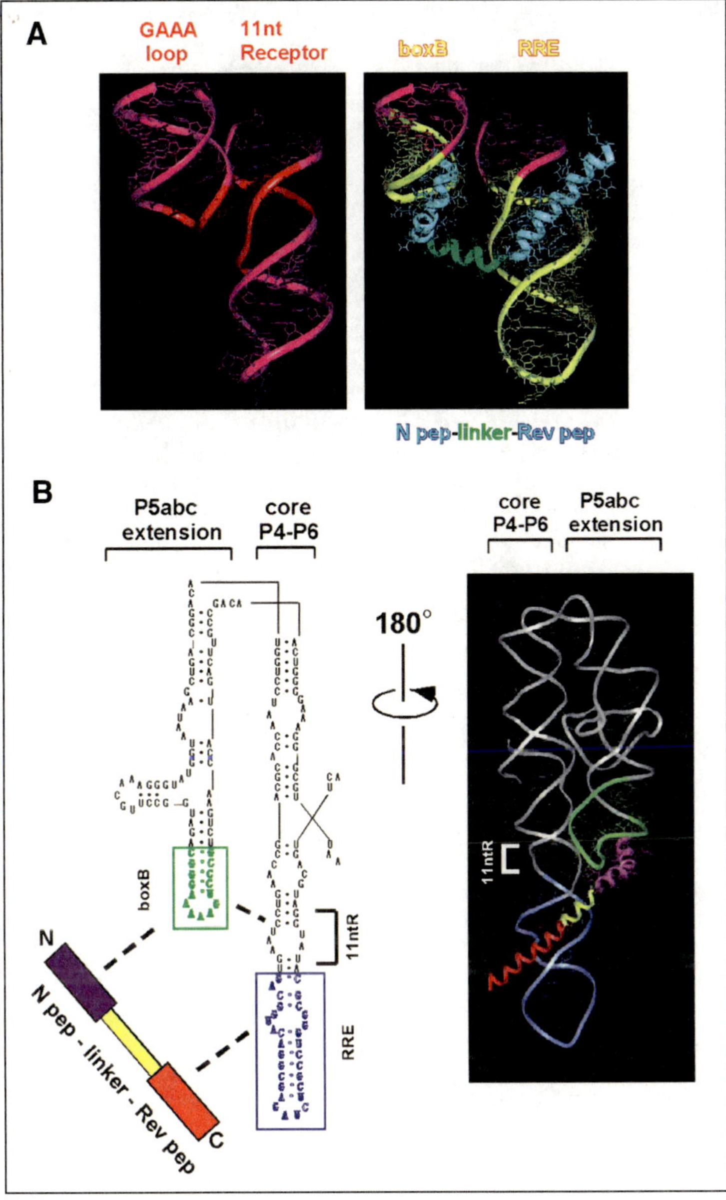

Figure 2. Molecular design of an RNA-protein complex based on the *Tetrahymena* ribozyme. A) Left, RNA-RNA interaction (red) between the P5b GAAA tetraloop and the 11-nt receptor within P6 in the crystal structure of the P4-P6 domain of the *Tetrahymena* ribozyme. Right, RNA-protein interactions between box B and N pep, and between RRE and Rev pep. In the designed model, RNA and peptide components are shown as yellow and light blue, respectively, and the linker region is shown as green. B) Alternative design of RNA-protein interactions based on the P4-P6 domain. Secondary and tertiary structures (left and right) of RNA-protein interactions between box B and N pep and between RRE and Rev pep. In the alternative model, the box B, N pep, RRE, and Rev pep are colored green, purple, blue and red, respectively. The linker region of the designed protein is shown as yellow. The structure on the right has been rotated 180° around a vertical axis relative to the structure on the left. Both RNA-RNA and RNA-protein interactions coexist in this model. (Reproduction from ref. 49 with permission.)

it is possible to develop a variety of artificial RNA-protein interactions in an allosterically regulated catalytic RNP.

Another molecular design for converting the *Tetrahymena* intron to a self-splicing RNP was attempted on the basis of high-resolution three-dimensional RNA and protein structures. In this study, molecules were designed to mimic a molecular evolution pathway from a ribozyme to a catalytic RNP.[49] One model molecule that mimicked a putative intermediary stage where RNA-RNA and RNA-protein interactions coexisted was more active than the parental *Tetrahymena* intron RNA,[49] suggesting that the association of a ribozyme with a protein might be advantageous for improving ribozyme activity (Fig. 2B). As expected, a newly designed protein that binds to a newly designed RNA facilitates reactivity in the absence of the original RNA-RNA interaction between the P5b tetraloop and the P6 receptor as described above. This demonstrates that molecular modeling serves as a versatile method for designing an RNP.[49]

Design of Protein-Dependent Allosteric Hammerhead Ribozymes

Rational molecular design is also applicable to the hammerhead ribozyme, which is a naturally occurring small ribozyme.[50-54] A protein-dependent hammerhead ribozyme was designed and constructed using the HIV-1 Rev protein (Fig. 3A).[55] Domain II of the hammerhead ribozyme was modified by fusing an RRE domain via a very short connection consisting of two base pairs. Presumably due to conformational alteration, the modified ribozyme is allosterically inactivated when HIV-1 Rev protein binds to the RRE. Ribozyme activity can be restored by the addition of an antibiotic that dissociates the protein from the RNA. Thus the concentration of the Rev protein and the corresponding antibiotic can be detected by monitoring ribozyme activity.

Alternatively, an inactive hammerhead ribozyme was constructed to use as a protein-dependent ribozyme by employing the Rev-RRE interaction (Fig. 3B).[55] The modified ribozyme possesses an additional oligonucleotide region that serves not only as an inhibitor that disrupts the binding of substrate RNA but also as a part of the binding site for the Rev protein. This design allows the substrate to bind to the ribozyme due to the displacement of the inhibitor region by the protein. In this design, the ribozyme is allosterically activated via a specific interaction between protein and RNA.

Analogously, a protein-binding domain for the protein kinase ERK2 was incorporated into the hammerhead ribozyme (Fig. 3C).[56,57] Fifty-fold activation of the modified ribozyme was achieved by adding the ERK2 kinase. As in the case of Rev-RRE, the inhibitor RNA strand is released from the ribozyme so that the specific binding of ERK2 establishes the active form. In another rational design, a newly incorporated protein-binding domain specific to the phosphorylated form of ERK2 allowed the construction of a hammerhead ribozyme activated specifically by phosphorylated ERK2. These ERK2-dependent ribozymes are competent both in vitro and in cell lysates, indicating that they can function as biosensors for detecting modified and unmodified proteins in signal transduction cascades.

Molecular design was also found to be applicable for allosteric regulation of the hammerhead ribozyme by employing the HIV-1 Tat-TAR interaction.[58] Ribozymes that can be modestly up-regulated or down-regulated by a Tat protein were constructed by installing a TAR motif into the ribozyme. An analogous design was successfully applied to construct a hammerhead ribozyme that is allosterically down-regulated by HIV-1 reverse transcriptase.[59]

Selection of Protein-Activated Artificial Ribozymes

An allosterically activated artificial ribozyme derived from the L1 ligase (Fig. 4A)[60] was attempted to be designed by employing a natural RNA-binding protein.[61] However, the L1 ligase ribozyme could not be converted to a protein-dependent ribozyme by installation of the

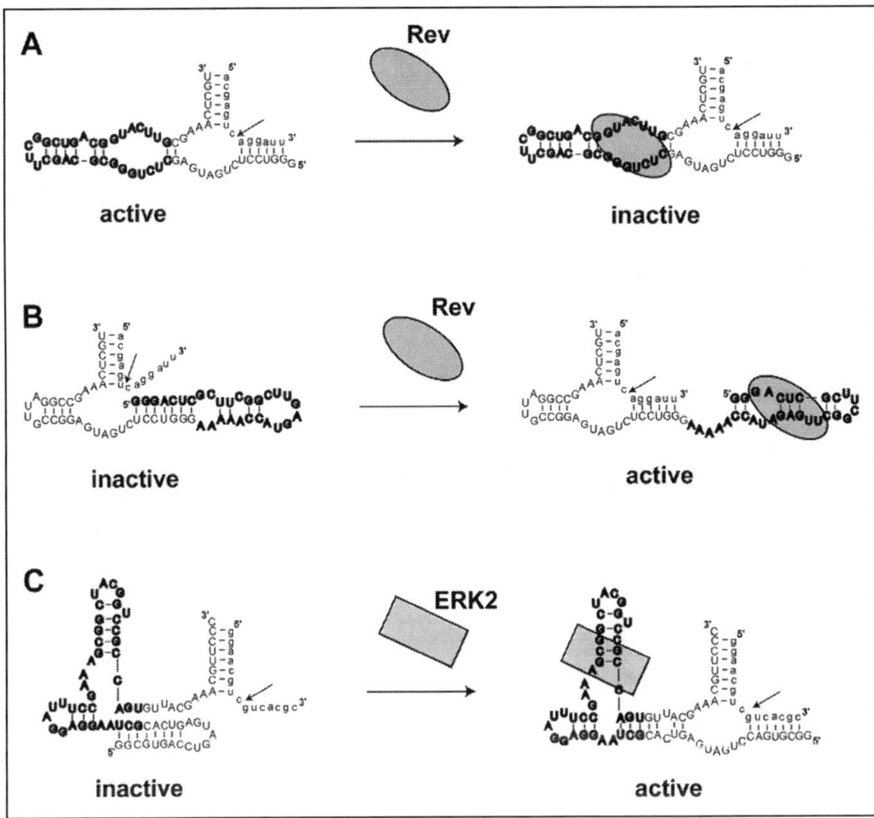

Figure 3. Allosteric control of the hammerhead ribozyme via an RNA-protein interaction. Cleavage sites are indicated with arrows, and sequences to be cleaved are shown with lowercase letters. A) The catalytically active ribozyme is inactivated by binding of Rev protein (grey circles) at the designed RNA binding site (bold font). B) An inactive ribozyme is activated by binding of the Rev protein (grey circles) at the designed binding site (bold font) in the RNA that induces an active conformation. C) An inactive ribozyme with additional stem-loop structure is activated by binding of the ERK2 protein (grey squares) at the designed RNA binding site (bold font).

corresponding protein-binding module, indicating that in this case protein binding was ineffective for converting the RNA structure into an alternative form. Because of this unsuccessful molecular design, new protein-binding domains responsible for activating the ribozyme were selected in vitro from a combinatorial library consisting of 50 random nucleotides, placed near the catalytic core of the RNA (Fig. 4B).[61] This enabled the construction of L1 derivatives for which activity is up-regulated by either CYT-18 or hen egg white lysozyme (Fig. 4C). Analogously, another peptide-dependent L1 ribozyme was selected in vitro from the same library (Fig. 4D).[62] This ribozyme was activated via a specific interaction with the HIV-1 Rev arginine-rich motif; the motif was recognized in the context of full-length HIV-1 Rev protein. These ribozymes have the potential to function as biosensors that specifically detect RNA-binding proteins. The simultaneous detection of many analytes has been achieved in an array format by immobilizing a variety of L1 ligase derivatives that can be regulated by specific RNA-binding proteins or other factors.[63]

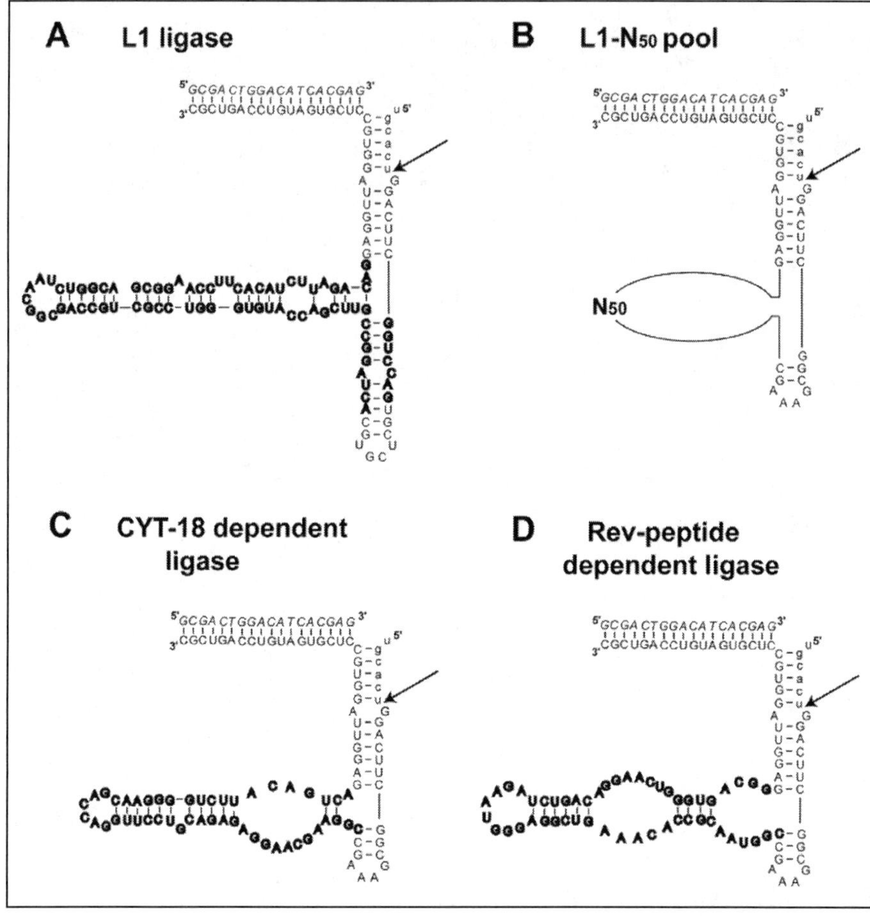

Figure 4. The protein-dependent L1 ligase ribozyme. A) Secondary structure of the L1 ligase ribozyme. B) Location of the N_{50} pool consisting of 50 random nucleotides that overlaps with a portion of the catalytic core of the L1 ligase. C,D) CYT-18-dependent and Rev-peptide-dependent L1 ligases with corresponding selected sequences. Nucleotides in bold font indicate the original L1 ligase sequence (A) or selected sequences (C,D). Ligation sites are indicated with arrows. The substrate RNA is shown with small letters. Guide DNA nucleotides essential for activation of the L1 ligase and its derivatives are shown in italics.

Applications of Protein-Switched Ribozymes

In conjunction with this expanding repertoire of protein-switched ribozymes, their use in biotechnology applications such as biosensing and gene therapy seems particularly attractive. Allosteric ribozymes regulated by small molecules have been demonstrated to function as biosensors.[53] Accordingly, the concentration of a target protein was successfully monitored with fluorescently labeled ribozymes.[55,63] A designed trans-splicing intron RNA can serve as a useful tool for repairing aberrant mRNA.[64] If these RNAs can be regulated by specific protein factors expressed in particular cells or tissues, the resulting protein-switched ribozymes could be employed for specific therapeutic purposes.

Perspective

In protein-switched ribozymes, specific RNA-binding proteins regulate RNA conformations to turn on or off ribozyme activity. Both in vitro and in vivo, this activity can be regulated with RNA-binding proteins in the case of natural ribozymes such as RNase P. Many experiments have demonstrated that natural and artificial catalytic RNAs are convertible to allosterically regulated ribozymes by practicing molecular design at the secondary or 3D structural levels or by performing selections from combinatorial libraries. The repertoire of RNA-protein combinations in these examples is currently restricted by the limited number of RNA motifs that are known to bind to protein. However, we expect that the number of such protein-binding motifs will soon increase because of rapid progress in the structural analyses of RNA-protein interactions in the ribosome and spliceosome, as well as those involving small noncoding RNAs. Moreover, in vitro selection techniques have been successfully applied for developing new artificial RNA-protein interactions.[57] Further advancements in the design and construction of protein-regulated ribozymes are expected by incorporating such newly identified interactions. Future studies along these lines are anticipated to delineate the rules governing the molecular evolution of functional RNPs in cells.

Acknowledgements

The work in our laboratories is supported by Grants-in-Aid for Scientific Research on Priority Areas (T. Inoue and Y. Ikawa) as well as the Takeda Science Foundation (T.I.), the Inamori and Iketani Foundations (Y.I.) and the Kyushu University Foundation (Y.I.).

References

1. Brion P, Westhof E. Hierarchy and dynamics of RNA folding. Annu Rev Biophys Biomol Struct 1997; 26:113-137.
2. Day-Storms JJ, Niranjanakumari S, Fierke CA. Ionic interactions between PRNA and P protein in Bacillus subtilis RNase P characterized using a magnetocapturebased assay. RNA 2004; 10:1595-608.
3. Herschlag D. RNA chaperones and the RNA folding problem. J Biol Chem 1995; 270:20871-20874.
4. Altman S, Kirsebom L, Talbot S. Recent studies of ribonuclease P. FASEB J 1993; 7:7-14.
5. Pace NR, Brown JW. Evolutionary perspective on the structure and function of ribonuclease P, a ribozyme. J Bacteriol 1995; 177:1919-1928.
6. Harris ME, Christian EL. Recent insights into the structure and function of the ribonucleoprotein enzyme ribonuclease P. Curr Opin Struct Biol 2003; 13:325-333.
7. Kazantsev AV, Krivenko AA, Harrington DJ et al. High-resolution structure of RNase P protein from Thermotoga maritima. Proc Natl Acad Sci USA 2003; 100:7497-7502.
8. Guerrier-Takada C, Gardiner K, Marsh T et al. The RNA moiety of ribonuclease P is the catalytic subunit of the enzyme. Cell 1983; 35:849-857.
9. Reich C, Olsen GJ, Pace B et al. Role of the protein moiety of ribonuclease P, a ribonucleoprotein enzyme. Science 1988; 239:178-181.
10. Pan T. Higher order folding and domain analysis of the ribozyme from Bacillus subtilis ribonuclease P. Biochemistry 1995; 34:902-909.
11. Loria A, Pan T. Domain structure of the ribozyme from eubacterial ribonuclease P. RNA 1996; 2:551-563.
12. Massire C, Jaeger L, Westhof E. Derivation of the three-dimensional architecture of bacterial ribonuclease P RNAs from comparative sequence analysis. J Mol Biol 1998; 279:773-793.
13. Waugh DS, Green CJ, Pace NR. The design and catalytic properties of a simplified ribonuclease P RNA. Science 1989; 244:1569-1571.
14. Haas ES, Banta AB, Harris JK et al. Structure and evolution of ribonuclease P RNA in Gram-positive bacteria. Nucleic Acids Res 1996; 24:4775-4782.
15. Krasilnikov AS, Yang X, Pan T et al. Crystal structure of the specificity domain of ribonuclease P. Nature 2003; 421:760-764.

16. Krasilnikov AS, Xiao Y, Pan T et al. Basis for structural diversity in homologous RNAs. Science 2004; 306:104-107.
17. Spitzfaden C, Nicholson N, Jones JJ et al. The structure of ribonuclease P protein from Staphylococcus aureus reveals a unique binding site for single-stranded RNA. J Mol Biol 2000; 295:105-115.
18. Stams T, Niranjanakumari S, Fierke CA et al. Ribonuclease P protein structure: Evolutionary origins in the translational apparatus. Science 1998; 280:752-755.
19. Crary SM, Niranjanakumari S, Fierke CA. The protein component of Bacillus subtilis ribonuclease P increases catalytic efficiency by enhancing interactions with the 5' leader sequence of pretRNA^{Asp}. Biochemistry 1998; 37:9409-9416.
20. Kurz JC, Niranjanakumari S, Fierke CA. Protein component of Bacillus subtilis RNase P specifically enhances the affinity for precursor-tRNA^{Asp}. Biochemistry 1998; 37:2393-2400.
21. Niranjanakumari S, Stams T, Crary SM et al. Protein component of the ribozyme ribonuclease P alters substrate recognition by directly contacting precursor tRNA. Proc Natl Acad Sci USA 1998; 95:15212-15217.
22. Green R, Noller HF. Ribosomes and translation. Annu Rev Biochem 1997; 66:679-716.
23. Ban N, Nissen P, Hansen J et al. The complete atomic structure of the large ribosomal subunit at 2.4 Å resolution. Science 2000; 289:905-920.
24. Nissen P, Hansen J, Ban N et al. The structural basis of ribosome activity in peptide bond synthesis. Science 2000; 289:920-930.
25. Moore PB, Steitz TA. The involvement of RNA in ribosome function. Nature 2002; 418:229-235.
26. Ogle JM, Carter AP, Ogle JM et al. Insights into the decoding mechanism from recent ribosome structures. Trends Biochem Sci 2003; 28:259-266.
27. Youngman EM, Brunelle JL, Kochaniak AB et al. The active site of the ribosome is composed of two layers of conserved nucleotides with distinct roles in peptide bond formation and peptide release. Cell 2004; 117:589-599.
28. Suzuki T, Terasaki M, Takemoto-Hori C et al. Structural compensation for the deficit of rRNA with proteins in the mammalian mitochondrial ribosome. Systematic analysis of protein components of the large ribosomal subunit from mammalian mitochondria. J Biol Chem 2001; 276:21724-21736.
29. O'Brien TW. Evolution of a protein-rich mitochondrial ribosome: Implications for human genetic disease. Gene 2002; 286:73-79.
30. Caprara MG, Lehnert V, Lambowitz AM et al. A tyrosyl-tRNA synthetase recognizes a conserved tRNA-like structural motif in the group I intron catalytic core. Cell 1996; 87:1135-1145.
31. Wallweber GJ, Mohr S, Rennard R et al. Characterization of Neurospora mitochondrial group I introns reveals different CYT-18 dependent and independent splicing strategies and an alternative 3' splice site for an intron ORF. RNA 1997; 3:114-131.
32. Mohr G, Caprara MG, Guo Q et al. A tyrosyl-tRNA synthetase can function similarly to an RNA structure in the Tetrahymena ribozyme. Nature 1994; 370:147-150.
33. Joyce GF, van der Horst G, Inoue T. Catalytic activity is retained in the Tetrahymena group I intron despite removal of the large extension of element P5. Nucleic Acids Res 1989; 17:7879-7889.
34. van der Horst G, Christian A, Inoue T. Reconstitution of a group I intron self-splicing reaction with an activator RNA. Proc Natl Acad Sci USA 1991; 88:184-188.
35. Atsumi S, Ikawa Y, Shiraishi H et al. Design and development of a catalytic ribonucleoprotein. EMBO J 2001; 20:5453-5460.
36. Murphy FL, Cech TR. GAAA tetraloop and conserved bulge stabilize tertiary structure of a group I intron domain. J Mol Biol 1994; 236:49-63.
37. Cate JH, Gooding AR, Podell E et al. Crystal structure of a group I ribozyme domain: Principles of RNA packing. Science 1996; 273:1678-1685.
38. Zarrinkar PP, Williamson JR. Kinetic intermediates in RNA folding. Science 1994; 265:918-924.
39. Sclavi B, Sullivan M, Chance MR et al. RNA folding at millisecond intervals by synchrotron hydroxyl radical footprinting. Science 1998; 279:1940-1943.
40. Young BT, Silverman SK. The GAAA tetraloop-receptor interaction contributes differentially to folding thermodynamics and kinetics for the P4-P6 RNA domain. Biochemistry 2002; 41:12271-12276.

41. Weiss MA, Narayana N. RNA recognition by arginine-rich peptide motifs. Biopolymers 1998; 48:167-180.
42. Patel DJ. Adaptive recognition in RNA complexes with peptides and protein modules. Curr Opin Struct Biol 1999; 9:74-87.
43. Battiste JL, Mao H, Rao NS et al. Alpha helix-RNA major groove recognition in an HIV-1 rev peptide-RRE RNA complex. Science 1996; 273:1547-1551.
44. Legault P, Li J, Mogridge J et al. NMR structure of the bacteriophage lambda N peptide/boxB RNA complex: Recognition of a GNRA fold by an arginine-rich motif. Cell 1998; 93:289-299.
45. Weeks KM. Protein-facilitated RNA folding. Curr Opin Struct Biol 1997; 7:336-342.
46. Weeks KM, Cech TR. Efficient protein-facilitated splicing of the yeast mitochondrial bI5 intron. Biochemistry 1995; 34:7728-7738.
47. Webb AE, Rose MA, Westhof E et al. Protein-dependent transition states for ribonucleoprotein assembly. J Mol Biol 2001; 309:1087-1100.
48. Atsumi S, Ikawa Y, Shiraishi H et al. Selections for constituting new RNA-protein interactions in catalytic RNP. Nucleic Acids Res 2003; 31:661-669.
49. Ikawa Y, Tsuda K, Matsumura S et al. Putative intermediary stages for the molecular evolution from a ribozyme to a catalytic RNP. Nucleic Acids Res 2003; 31:1488-1496.
50. Hammann C, Lilley DM. Folding and activity of the hammerhead ribozyme. ChemBioChem 2002; 3:690-700.
51. Vaish NK, Kore AR, Eckstein F. Recent developments in the hammerhead ribozyme field. Nucleic Acids Res 1998; 26:5237-5242.
52. Soukup GA, Breaker RR. Nucleic acid molecular switches. Trends Biotechnol 1999; 17:469-476.
53. Breaker RR. Engineered allosteric ribozymes as biosensor components. Curr Opin Biotechnol 2002; 13:31-39.
54. Silverman SK. Rube Goldberg goes (ribo)nuclear? Molecular switches and sensors made from RNA. RNA 2003; 9:377-383.
55. Hartig JS, Najafi-Shoushtari SH, Grune I et al. Protein-dependent ribozymes report molecular interactions in real time. Nat Biotechnol 2002; 20:717-722.
56. Vaish NK, Dong F, Andrews L et al. Monitoring post-translational modification of proteins with allosteric ribozymes. Nat Biotechnol 2002; 20:810-815.
57. Vaish NK, Kossen K, Andrews LE et al. Monitoring protein modification with allosteric ribozymes. Methods 2004; 32:428-436.
58. Wang DY, Sen D. Rationally designed allosteric variants of hammerhead ribozymes responsive to the HIV-1 Tat protein. Comb Chem High Throughput Screen 2002; 5:301-312.
59. Hartig JS, Famulok M. Reporter ribozymes for real-time analysis of domain-specific interactions in biomolecules: HIV-1 reverse transcriptase and the primer-template complex. Angew Chem Int Ed 2002; 41:4263-4266.
60. Robertson MP, Ellington AD. In vitro selection of an allosteric ribozyme that transduces analytes to amplicons. Nat Biotechnol 1999; 17:62-66.
61. Robertson MP, Ellington AD. In vitro selection of nucleoprotein enzymes. Nat Biotechnol 2001; 19:650-655.
62. Robertson MP, Knudsen SM, Ellington AD. In vitro selection of ribozymes dependent on peptides for activity. RNA 2004; 10:114-127.
63. Hesselberth JR, Robertson MP, Knudsen SM et al. Simultaneous detection of diverse analytes with an aptazyme ligase array. Anal Biochem 2003; 312:106-112.
64. Long MB, Jones IIIrd JP, Sullenger BA et al. Ribozyme-mediated revision of RNA and DNA. J Clin Invest 2003; 112:312-318.

CHAPTER 4

Fluorescence-Signaling Nucleic Acid-Based Sensors

Razvan Nutiu, Lieven P. Billen and Yingfu Li*

Abstract

It is widely known that two single-stranded nucleic acids with complementary sequences have the inherent ability to form Watson-Crick duplex structures. The simplicity and sequence-specificity of duplex structure formation, the high chemical stability of a duplex, and the convenience of automated synthesis have made DNA oligonucleotides an ideal choice as probes for the detection of nucleic acids. Recently developed in vitro selection techniques permit creation of DNA and RNA "aptamers" that are capable of binding a wide variety of nonnucleic acid targets with high affinity and specificity. Aptamers have considerably broadened the utility of nucleic acids as probes for detection of biological and nonbiological targets. In vitro selection also allows generation of artificial ribozymes (catalytic RNAs) and deoxyribozymes (catalytic DNAs) with desirable functions. Aptamers, ribozymes, and deoxyribozymes have become increasingly valuable molecular tools in the form of switches and sensors. Unfortunately, binding or catalytic actions by these switches and sensors do not usually lead to an easily detectable signal, and the lack of a facile reporting method could substantially reduce their value. To facilitate the exploitation of nucleic acid switches and sensors for detection-related applications, many recent studies have explored fluorescence signaling as a convenient approach for the reporting of binding and catalytic events. This chapter is devoted to the discussion of these efforts. The reporter molecules to be described include molecular beacons, signaling aptamers, and signaling ribozymes and deoxyribozymes.

Introduction

Since the report of the first biosensor in 1967,[1,2] tremendous progress has been made in the biosensor field. By conventional definition, a biosensor refers to an analytical device consisting of two important components: a molecular recognition element (MRE) for target (analyte) detection and a transducer that physically reports the MRE-analyte interaction. In such a device, the MRE acts as an affinity probe that specifically seeks an analyte from a complex biological sample for binding. Many naturally occurring and artificial substances can be explored as MREs, and the choice of a specific MRE for a particular application is often dependent on the nature of the analyte and the availability of MREs.[3] The widely

*Corresponding Author: Yingfu Li—Department of Biochemistry, McMaster University, 1200 Main Street West, Hamilton, Ontario L8N 3Z5, Canada. Email: liying@mcmaster.ca

Nucleic Acid Switches and Sensors, edited by Scott K. Silverman. ©2006 Landes Bioscience and Springer Science+Business Media.

utilized MREs include natural receptors,[4] protein enzymes and antibodies,[5] cells,[6] organelles,[7] tissues[8] and molecularly imprinted polymers.[9]

The suitability of nucleic acids as specific MREs for the detection of DNA and RNA targets is clear due to their inherent ability to form Watson-Crick duplexes. In a broad sense, Southern blotting[10] can be regarded as a primitive biosensor that uses nucleic acids as MREs. Over the last decade, the exploitation of nucleic acids as MREs has been significantly extended beyond the detection of nucleic acid targets, thanks to the ground-breaking discoveries that certain single-stranded DNA or RNA molecules termed "aptamers" can form defined tertiary structures for binding nonnucleic acid targets.[11,12] From the extensive studies reported to date, it is increasingly apparent that DNA and RNA aptamers represent a new class of versatile MREs for the detection of a broad scope of nonnucleic acid analytes, including proteins and metabolites.[13,14]

Aptamers offer several distinct advantages as MREs. Aptamers are isolated in vitro, and the targets for aptamer creation can be any compounds including toxins. Once an aptamer is identified, it can be prepared either by automated DNA synthesis at very low cost and with high batch-consistency (in the case of DNA) or by simple in vitro transcription (in the case of RNA). Moreover, it is easy to modify aptamers to introduce various reactive groups, affinity tags, or reporting moieties that are required for biosensing applications. Finally, DNA aptamers have a long shelf life and usually retain activity following a denaturation-renaturation process.

Of course, molecular recognition elements can only be formulated into a biosensing device if the MRE-analyte recognition event can be transduced into an easily detectable signal. One important step in developing biosensing systems that exploit aptamers as MREs is to establish effective methods that report an aptamer-analyte binding event. Signal transduction can be achieved through electrochemical, mechanical, piezoelectric, or fluorescent means, and each of these methods have been applied to aptamers.[15-18] Among these methods, fluorescence signaling is highly desirable because of the ease of detection, the availability of diverse measurement methods, and the existence of a diverse collection of fluorophores and quenchers suitable for nucleic acid modification.[19] In comparison to more traditional radiolabeling approaches that are often exploited in nucleic acid assays, fluorescence detection confers several advantages. First, fluorescent substrates are not subjected to the signal decay inherent to radioactive substrates. Second, fluorescence is amenable to real-time or continuous assays, in contrast to the laborious and time-consuming discontinuous assays necessary when working with radioactive substrates. Third, fluorescence-based platforms lend themselves well to preexisting fiber-optic and DNA microarray technologies.[20,21] Therefore, various strategies for the design of effective fluorescence-signaling aptamers are being actively pursued by many research groups.

Similar to aptamers, catalytic nucleic acids (ribozymes and deoxyribozymes) can be generated de novo by in vitro selection from an extraordinarily large population of single-stranded RNA or DNA molecules.[22] Although no catalytic DNA has been found in nature, many research laboratories around the world have isolated hundreds of DNA sequences that facilitate more than a dozen chemical transformations, often involving nucleic acid substrates.[23-26] For example, DNA enzymes have been isolated that cleave DNA[27] and RNA,[28,29] ligate RNA[30] and phosphorylate DNA.[31] Ribozymes and deoxyribozymes may have limited value for direct application as MREs because of the relatively small scope of substrates and/or cofactors. However, the recent development of allosteric nucleic acid enzymes—RNA and DNA molecules consisting of both an aptamer domain and a catalytic domain—has made catalytic RNA and DNA very useful as part of MREs.[32] As a result, increasing efforts have been focused on developing fluorescence-based sensor molecules made of ribozymes and deoxyribozymes.

In this chapter, we discuss recent progress in the development of novel fluorescent nucleic acid probes as MREs for the detection of nucleic acid targets and nonnucleic acid targets. We first describe molecular beacons—hairpin-shaped fluorescent DNA probes that are extremely useful for nucleic acid detection. This is followed by a review of the engineering of signaling aptamers. Finally, we discuss examples of engineering fluorescence-signaling ribozymes and deoxyribozymes identified either by in vitro selection or by post-selection modifications.

Molecular Beacons for Nucleic Acid Detection

Because standard nucleic acid bases are inherently nonfluorescent, chemical modifications are required to make them fluorescent. The simplest fluorescence-signaling method for nucleic acid detection is the use of an organic chromophore that can bind to nucleic acids with accompanying changes in its fluorescence properties. Many compounds such as ethidium bromide and Hoechst 33342 have very weak intrinsic fluorescence in solution but display significantly enhanced emission when bound to duplex DNA or RNA. Nucleic acids can also be made fluorescent by the use of nucleobase analogs such as 2-aminopurine (for adenine) or isoxanthopterin (for guanine). The most attractive method for making DNA or RNA fluorescent is usually the covalent attachment of fluorophores or quenchers directly onto the nucleic acid. An external fluorophore can be joined either to an end of a nucleic acid chain or coupled onto an internal nucleotide. The modification can be carried out during the solid-phase synthesis of DNA or RNA using fluorophore-containing phosphoramidite monomers; during the enzymatic synthesis of DNA or RNA using fluorophore-modified nucleoside triphosphates; or through post-synthetic coupling using facile reactions and simple procedures.

A new group of fluorescent DNA probes, known as molecular beacons (MB), have become increasingly popular. MBs are single-stranded DNA probes doubly modified with a fluorophore (F) and a quencher (Q) at the two ends of the DNA strand and are useful for DNA or RNA detection with a signaling mechanism that is illustrated in Figure 1.[33] The sequence of each MB is designed to form a stem-loop structure. The stem of a MB is formed between two complementary segments of usually 5-7 nucleotides (nt) in length that are located at the opposite ends of the DNA strand. The loop sequence is specifically designed to be complementary to the sequence of a target DNA or RNA molecule. Because of these design elements, an MB can adopt two structural states, the "open" state and the "closed" state. In the closed state, the fluorophore is located in close proximity to the quencher, and

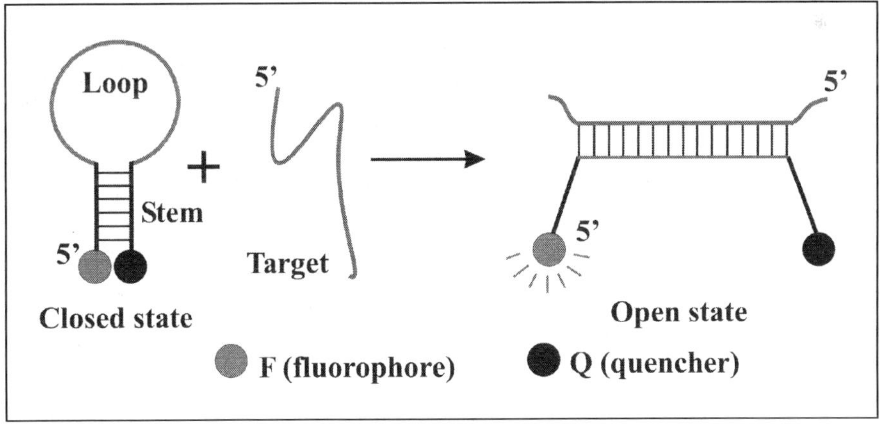

Figure 1. Standard molecular beacons. A fluorophore (F) and a quencher (Q) are covalently linked to the two ends of the hairpin-shaped DNA probe. The formation of loop-target duplex structure causes the separation of F from Q and the generation of a fluorescence signal. This figure is adapted from reference 39.

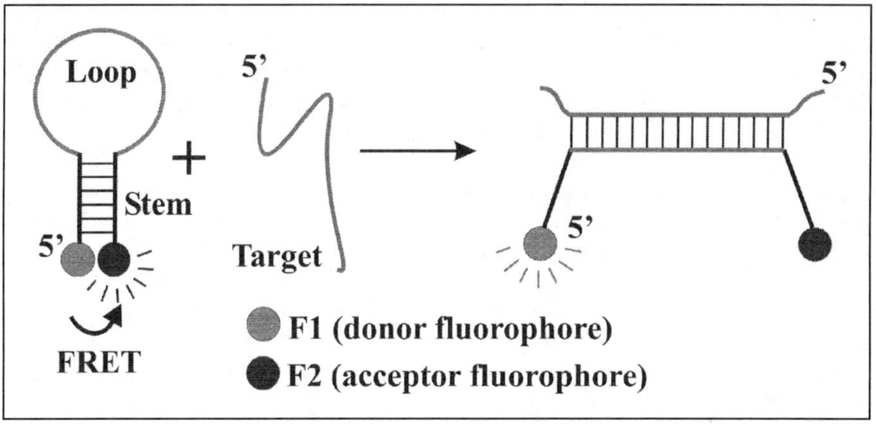

Figure 2. Molecular beacons with two fluorophores. These molecular beacons offer the options of monitoring the disappearance of FRET from the acceptor fluorophore and appearance of fluorescence signal from the donor fluorophore.

fluorescence is not observed. However, when the target is present, the MB adopts the open state due to the formation of more stable loop-target duplex. Because in the open state the quencher is spatially separated from the fluorophore, the emission of fluorescence is restored. The loop size and stem strength of an MB can be fine-tuned such that only a target with a fully complementary sequence can efficiently induce the transition from the closed state to the open state. Therefore, MBs offer very high sequence specificity in target recognition and can usually discriminate sequences that differ by a single nucleotide. Because MBs have a built-in fluorescent switch, target detection can be performed in real time and does not require a lengthy procedure to separate the target-probe complex from free probes. Furthermore, several MBs labeled with different fluorophores can be used to conduct a multiplexed assay for the analysis of different targets in one solution.[34] The fluorescence reporting of MBs is very sensitive, and fluorescence enhancements upon target binding of up to two orders of magnitude have been reported.[35]

MBs can be designed or utilized in several variations, each offering advantages for particular applications. For example, dual-fluorophore labeled MBs offer the option of monitoring emission intensities at two different wavelengths, indicative of engagement or disengagement of FRET (fluorescence resonance energy transfer) between the two fluorophores (Fig. 2).[36] The two beacons/one target FRET approach (Fig. 3) reduces the chance of false positives that may result from factors other than target-probe duplex formation.[37] The two MBs are designed to target the neighboring sequences of the same DNA or RNA molecule. The fluorophores on the two MBs are chosen for their ability to engage each other for FRET. When both MBs are opened by the target, the two fluorophores are brought in close proximity to activate FRET. However, if MBs are open due to nonspecific interactions or degraded by a contaminating nuclease, FRET is absent because the two fluorophores are not located near each other.

Wavelength-shifting MBs (wsMBs; Fig. 4) are triply labeled with two fluorophores and one quencher.[38] The two fluorophores are chosen to be a FRET pair; therefore the excitation of the first fluorophore (called the harvester) will lead to the emission of the second fluorophore (emitter), provided that the quencher is physically distant due to target binding. wsMBs offer enhanced detection sensitivity relative to the simpler approach of Figure 2 due to an increased Stokes shift. Most attractively, several wsMBs can be constructed with a common harvester and different emitters for a multiplexed assay that uses monochromatic excitation light.

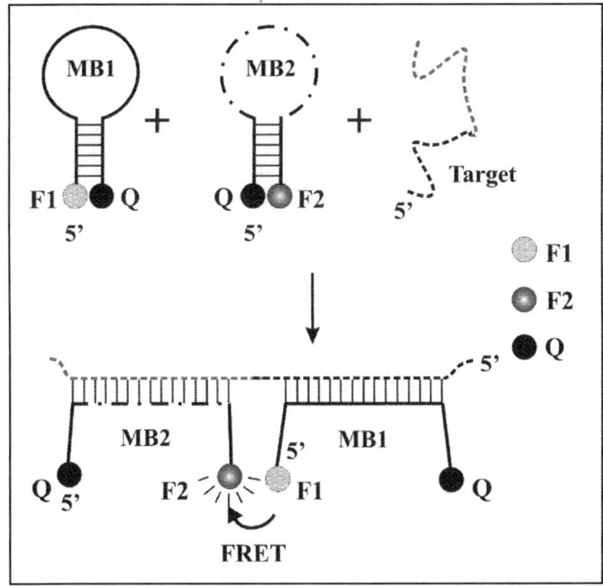

Figure 3. Two-MB/one-target design. This approach enhances the detection specificity and reduces false positive signaling.

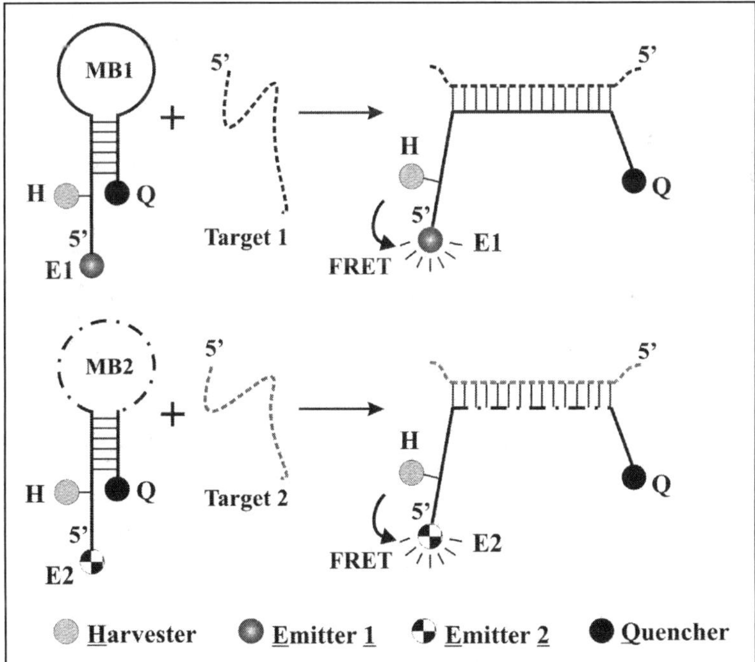

Figure 4. Wavelength-shifting molecular beacons. These molecular beacons can be used for multiplexed assays using a fixed excitation wavelength, if two (or more) beacons are used for two (or more) targets.

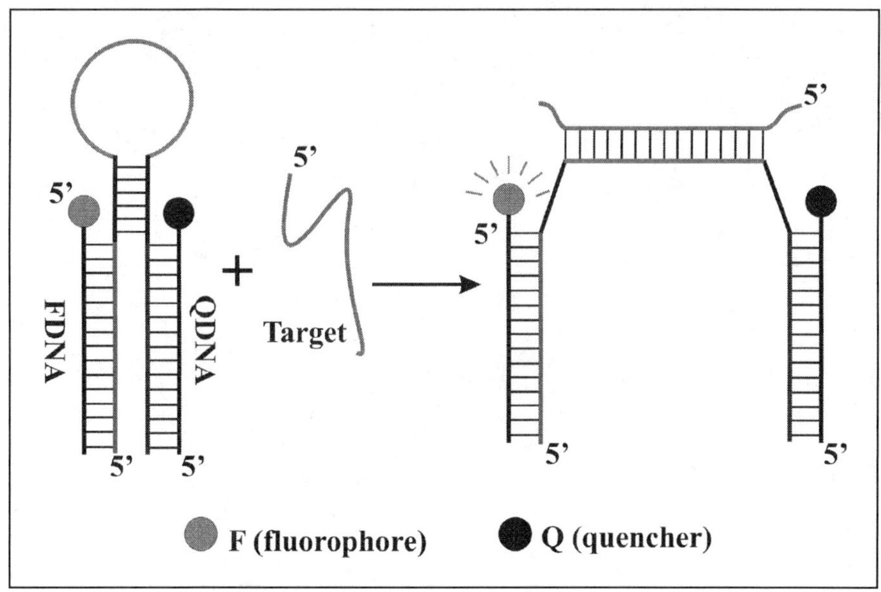

Figure 5. Tripartite molecular beacons. F and Q are linked to two separate short oligonucleotides (denoted FDNA and QDNA), which hybridize with corresponding single-stranded arms extended beyond the short hairpin stem. This figure is adapted from reference 39.

A standard MB can also be reformulated into a tripartite molecular beacon (TMB) assembly (Fig. 5).[39] In TMBs, the fluorophore and the quencher are not covalently attached to the ends of the hairpin probe; instead they are linked to the ends of two new short oligonucleotides denoted FDNA (containing a fluorophore) and QDNA (having a quencher). FDNA and QDNA are designed to hybridize with the two extended arms of the unmodified hairpin probe and convert it into an MB-like assembly (Fig. 5). Since a FDNA/QDNA combination can be used as a universal pair of labeling probes, the TMB approach is particularly attractive for applications that require a large number of MBs. In other words, the FDNA/QDNA pair can be used universally with many MBs since they (and their complementary sequences) interact with neither the stem that closes the MB nor the target recognition sequence.

Ideally, a MB should produce a large fluorescence signal upon target binding. Because the signal generation is a de-quenching process in the case of Figure 1, an appropriate quencher must be chosen for a particular fluorophore to reduce the level of background fluorescence (i.e., the fluorescence intensity prior to the target addition). Some small organic molecules such as DABCYL and Black Hole Quenchers are popular due to their high quenching efficiency for most fluorophores, as well as their thermal stability and photo-stability. Other nonorganic quenchers have also been described. For example, a 1.4-nm diameter gold nanoparticle is a highly effective quencher for many fluorophores in the MB context.[40] Gold can also be used as a surface for DNA immobilization, and in such a setting there is no need for a separate quencher, because the gold surface acts as a quencher.[41] The main drawback of nanogold is its thermal instability, making it impossible to use such a quencher in applications that involve high-temperature steps, such as the polymerase chain reaction (PCR).

MBs have been studied for many nucleic acid detection and reporting applications. For example, MBs have widely been used as DNA-reporting probes for PCR. A PCR cycle consists of three temperature steps: denaturation (94°C), annealing (50-60°C) and extension (72°C). MBs are usually designed to function during the annealing step. When the DNA amplification target (amplicon) is present in low amounts, most MB molecules stay in the closed state and emit low fluorescence. When the amplicon concentration increases during PCR, more MB molecules bind to their targets and become highly fluorescent. The use of MBs as reporter molecules in PCR offers two attractive features. First, the monitoring of DNA amplicons can be achieved in real time without any post-PCR sample manipulation. Second, MB-based detection methods are highly sequence-specific because MBs will only produce a fluorescence signal when the amplicon contains the sequence complementary to the loop segment of the hairpin structure.

One important application for MBs is as PCR probes for human allele genotyping. For example, MBs have been used as elegant tools to screen the human population for susceptibility to HIV-1 infection. It is known that individuals with a 32-nt deletion in the β-chemokine receptor 5 gene in homozygous chromosomes (mutant homozygous), with a 32-nt deletion in only one chromosome (mutant heterozygous), and with no deletion (wild-type homozygous) are largely resistant, partially resistant, and susceptible to HIV-1 infection, respectively. Kostrikis et al exploited this fact in the design of the two MBs shown in Figure 6. The first MB (MB1, labeled with a fluorophore that emits green fluorescence) targets the deletion sequence for binding, while the second MB (MB2, labeled with a fluorophore that emits red fluorescence) targets the two stretches of nucleotides sandwiching the deletion sequence for binding.[42] When the DNA from an individual is used as a PCR template and the two MBs are used as detection probes, increases in fluorescence from MB1 only, from MB2 only, and from both MB1 and MB2 indicate wild-type homozygous, mutant homozygous, and mutant heterozygous, respectively.

The power of PCR, coupled with real-time and sequence-specific detection by MBs, allows the fast and accurate characterization of minute amounts of biologically relevant nucleic acid material. To date, MBs have been extensively used for DNA, RNA and pathogenic detection; many examples can be found at http://www.molecularbeacons.com/Publications.html. MBs have also been exploited in bioanalytical assays designed to characterize DNA- or RNA-binding proteins;[43] report enzymatic processes such as cleavage of DNA or RNA by nucleases;[44,45] monitor in vitro transcription;[46] and report on rolling circle amplification.[47] Moreover, as we will discuss next, the molecular beacon concept has been adapted to the design of signaling aptamers and signaling catalytic nucleic acids.

Signaling Aptamers

The development of in vitro selection techniques has led to the generation of DNA and RNA aptamers that bind a wide variety of nonnucleic acid targets that include proteins and metabolites. Like any other nucleic acid molecules, aptamers are inherently nonfluorescent, and therefore post-selection modifications with external fluorophores must be performed to convert aptamers into fluorescence-signaling reporters.

An ideal rational design strategy for engineering signaling aptamers should have three key features. First, the method should be easy to apply to any given aptamer regardless of its size and structural properties. This is important because aptamers have variable sizes and vastly different secondary and tertiary structures. Some aptamers do not have an easily determinable secondary structure, and the tertiary structures of most aptamers are not readily available. Second, the method should be capable of designing signaling aptamers with large signaling magnitudes, fast responses, and real-time signaling capabilities. Signaling aptamers

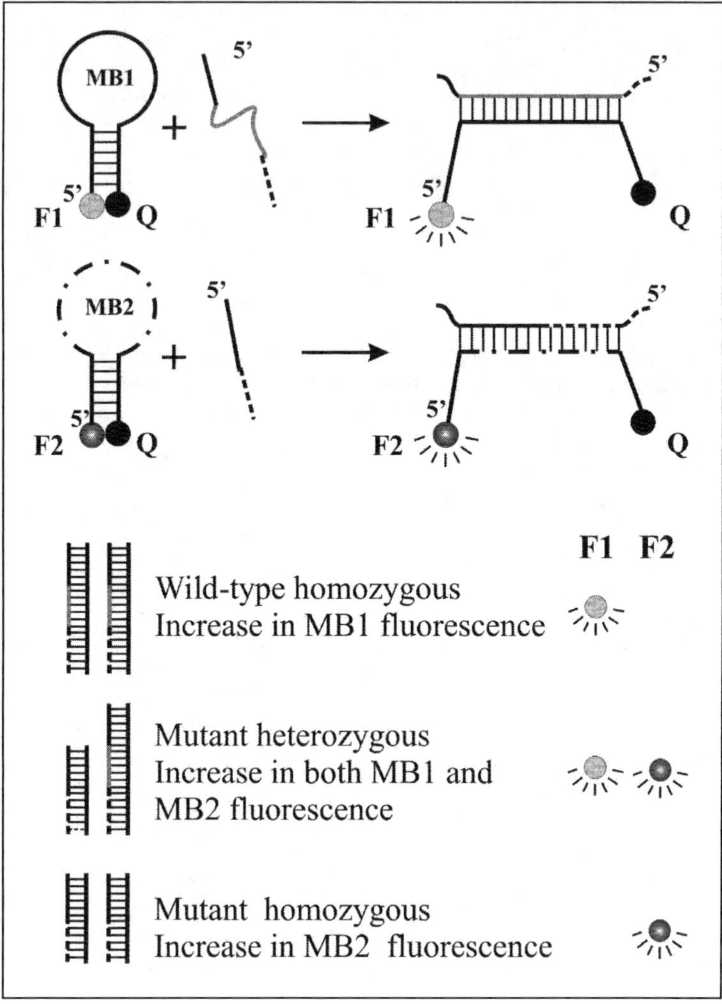

Figure 6. Molecular beacons for genotyping. A pair of molecular beacons labeled with two different fluorophores permits identification of alleles with or without deletion mutations.

that exhibit large fluorescence enhancements upon target binding increase the sensitivity and accuracy of the corresponding assays. The real-time reporting capability allows rapid sample measurements and permits demanding applications such as high-throughput screening. Third, the method should be adaptable for the design of a signaling aptamer from any existing aptamer without altering its affinity and specificity.

One signaling aptamer design strategy is to attach a single fluorophore at different locations on the aptamer and test each construct for altered spectroscopic properties upon target binding. Jhaveri et al first reported this approach in 2000 (Fig. 7).[48] They used two existing ATP aptamers—one RNA and one DNA—with defined tertiary structures as model examples. Several aptamers, each with a fluorophore attached at a location distant from the target-binding site, were examined. Five modified aptamers did not show fluorescence changes in response to

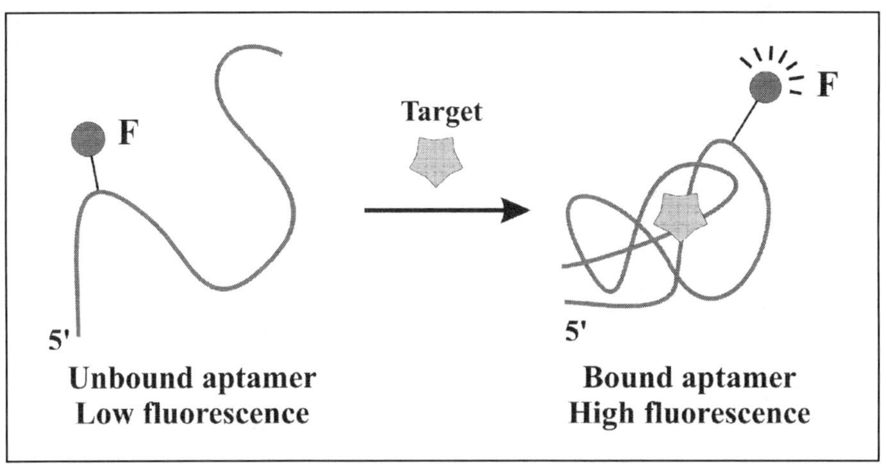

Figure 7. Single-fluorophore signaling aptamers by rational design. Labeling an aptamer with a fluorophore at a location where a substantial structural reorganization upon target binding induces a change in the fluorescence property of the attached fluorophore. This figure is adapted from reference 86.

the addition of ATP, and the remaining two registered a 25-45% increase in fluorescence intensity upon ATP addition. The failures might have been caused by disruption of the correct folding of the aptamer as a result of fluorophore attachment, or by minimal alteration of the emission of the attached fluorophore when the aptamer binds to the target molecule. Aptamers with improved signaling magnitudes were reported in a later study where bis-pyrene was used as the fluorescent tag.[49]

Jhaveri et al also investigated the direct acquisition of signaling aptamers through in vitro selection (Fig. 8). To do so, they constructed a pool of random-sequence RNAs each containing one or a few fluorescein-modified uridines.[50] After several rounds of selective amplification using ATP as the binding target, a few aptamers that act as real-time reporters for ATP were obtained. The best signaling aptamer exhibited approximately two-fold fluorescence intensity increase at the saturating ATP concentration. Interestingly, the fluorescein label on the aptamer could be substituted by other fluorophores without affecting the aptamer's target-binding affinity and specificity. It is noteworthy that their selection also generated several aptamers that failed to register significant fluorescence enhancement upon target binding, suggesting that even the selection approach offers no guarantee regarding the generation of aptamers that are effective fluorescent reporters.

As discussed earlier, MBs usually exhibit large signaling magnitudes upon target binding. Although standard molecular beacons are only useful for nucleic acid detection, several groups have attempted to adapt the same principle for the design of signaling aptamers. The resulting constructs are often termed aptamer beacons. Hamaguchi et al made the most straightforward adaptation (Fig. 9A) by adding a few nucleotides onto the 5'-end of a small thrombin DNA aptamer to engage the 3'-end of the aptamer into a hairpin structure.[51] When the protein target is absent, the aptamer beacon forms the closed-state structure in which fluorescence is quenched. In the presence of thrombin, the aptamer beacon forms the target-complexed structure in which the fluorophore and the quencher are spatially separated, resulting in a larger fluorescence signal. A two-chain aptamer beacon approach has also been reported.[52] In this design, Yamamoto et al split an HIV Tat protein-binding RNA aptamer into two molecules, one of which was formulated into a molecular beacon (Fig. 9B).

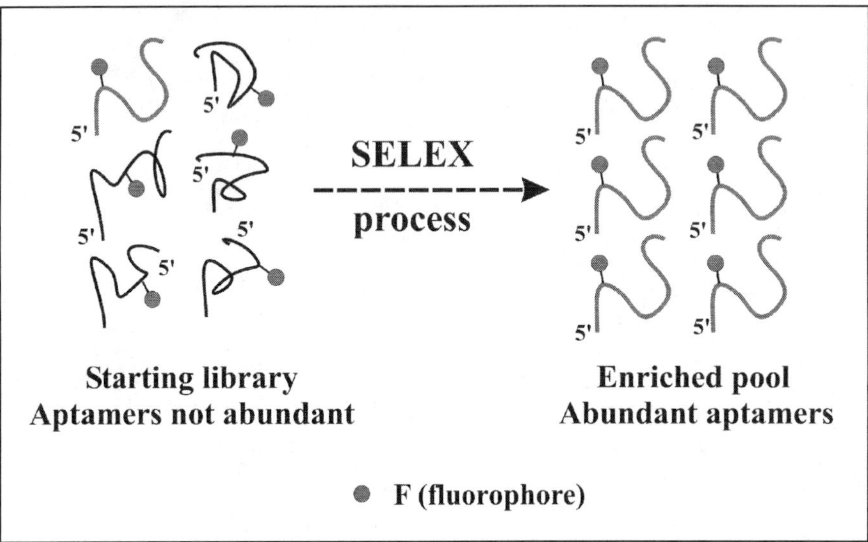

Figure 8. Single-fluorophore signaling aptamers by in vitro selection. Signaling aptamers are generated de novo by SELEX using a random-sequence DNA or RNA library in which each DNA or RNA molecule is labeled with one or a few fluorophores. This figure is adapted from reference 86.

In the absence of Tat, the two RNA molecules do not interact, and the solution has low fluorescence intensity. When Tat is introduced, the molecular beacon and the other half of the aptamer assemble into a single unit for Tat binding. The dissociation of the hairpin structure is accompanied by an increase of fluorescence intensity.

Other groups have explored fluorescence quenching as an alternate way to track target-aptamer recognitions, where the detection depends on loss (rather than gain) of fluorescence signal upon binding of the target to the aptamer. Stojanovic et al used a two-chain-assembly approach in the design of two aptamer reporters, one that binds cocaine and the other for ATP recognition (Fig. 9C).[53] In the absence of the target, the two chains of the aptamer (one labeled with a quencher and the other with a fluorophore) do not associate strongly, and therefore the fluorescence intensity of the solution is high. Introduction of the target promotes the assembly of the two chains for target binding, resulting in fluorescence quenching. Moreover, the two aptamer reporters, each labeled with a different fluorophore, can function to some extent in the same solution, suggesting the development of aptamer-based multiplexed detection. Another fluorescence-quenching thrombin signaling aptamer, as illustrated in Figure 10, was later described by Li et al.[54] It is known that the two ends of the anti-thrombin DNA aptamer are located next to each other in the folded guanine-quartet complex structure. Based on this folding property, Li et al designed a signaling aptamer by appending a fluorophore-quencher pair to the two ends of the aptamer sequence.[54] The resulting aptamer was able to perform real-time reporting of thrombin by fluorescence quenching. Similarly, a DNA aptamer for the B chain of platelet-derived growth factor (PDGF) labeled with a fluorophore-quencher pair at its termini shows fluorescence quenching upon addition of PDGF.[55]

Our group has recently described a different approach for the design of signaling aptamers by exploiting the common ability of each aptamer to form two completely different structural forms: a relatively weak duplex structure with a complementary DNA oligonucleotide and a stronger complexed structure with the selected target (Fig. 11). First, we attach a fluorophore onto the aptamer and a quencher onto the complementary oligonucleotide that

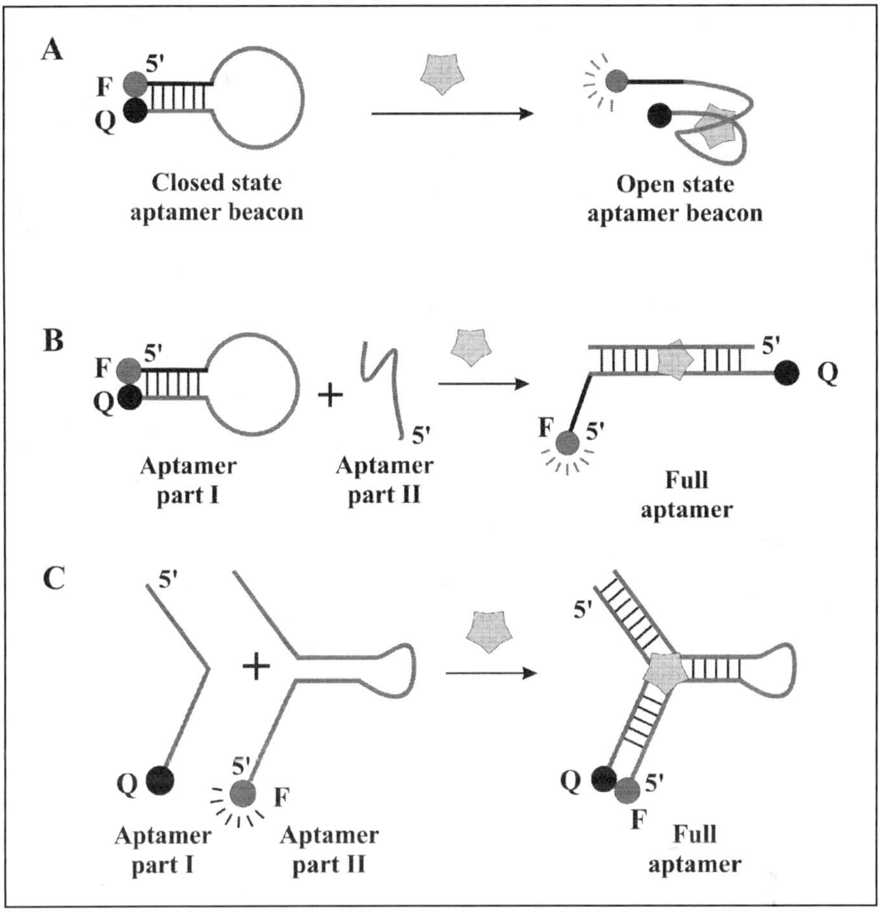

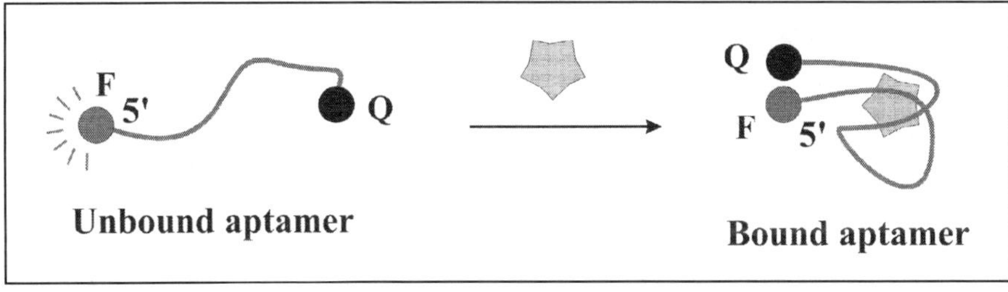

Figure 9. A) Signaling aptamers by one-chain molecular beacon approach. An aptamer sequence is extended so that it can be formulated into a molecular beacon. B) Signaling aptamers by two-chain molecular beacon approach. An aptamer is split into two chains, one of which is formulated into a molecular beacon. C) Signaling aptamers by two-chain molecular assembly. An aptamer is split into two chains, each of which is labeled with a fluorophore or a quencher. This figure is adapted from reference 86.

Figure 10. Signaling aptamers based on tertiary folding. This strategy works with aptamers whose tertiary folding brings the two ends of the aptamer chain into close proximity. This figure is adapted from reference 86.

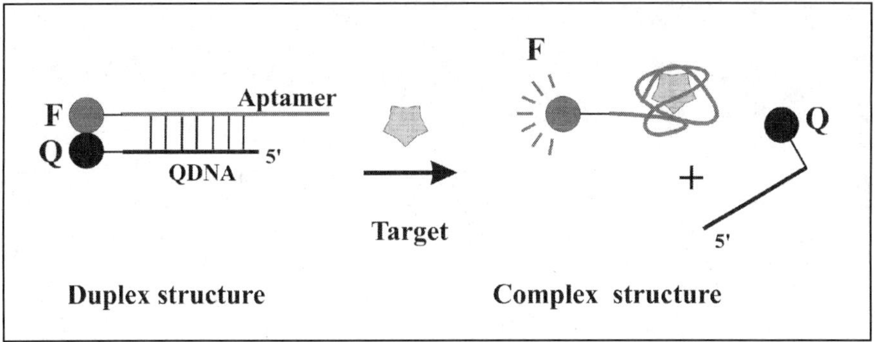

Figure 11. Structure-switching signaling aptamers. A modified aptamer is configured into a low-fluorescence duplex structure state with the fluorophore-containing aptamer and quencher-bearing QDNA. Target binding releases QDNA, leading to a highly fluorescent complexed structure. This figure is adapted from reference 86.

can anneal with part of the aptamer sequence. In the absence of the target, the oligonucleotides form a low-fluorescence double-helix structure. When the target is present, the aptamer switches from the duplex structure to the complexed structure, releasing the quencher-containing antisense oligonucleotide into solution. Spatial separation of the fluorophore and quencher generates a highly intense fluorescent signal. We called these designed signaling aptamers "structure-switching signaling aptamers" to emphasize the common structure-switching mechanism. Two aptamers, one for ATP and one for thrombin, with considerably different sizes, tertiary structures, and affinities for their targets, were successfully transformed into structure-switching signaling aptamers.[56]

Recently, a novel fluorophore-tagging technique has been described.[57] Fluorescamine (FCM) is a fluorogenic compound that is known to react very rapidly with amine groups. If an aptamer is synthesized with one or a few 2'-amine substituted nucleotides, formation of the usually compact aptamer-target complex structure will reduce the FCM-tagging activity of the amine groups. Several variants of the anti-ATP DNA aptamer with 2'-amine substituted nucleotides were constructed and tested for FCM-tagging activities in the presence of and absence of ATP. One construct showed considerably reduced reactivity when bound to ATP. This construct was then transformed into a signaling aptamer by appending a new fluorophore (F1) on the end of the DNA sequence capable of engaging FCM (F2) for FRET (Fig. 12). In the presence of ATP, FCM was not able to react efficiently with the 2' amine group on the aptamer, and therefore no FRET was observed.

Signaling Ribozymes and Deoxyribozymes

Many signaling ribozymes and deoxyribozymes have also been described in recent years. Fluorescently labeled ribozymes and deoxyribozymes have two significant advantages relative to other signaling approaches. First, the catalytic activity of a signaling enzyme with a built-in fluorescent switch can be followed conveniently by the change of fluorescence intensity in real time. If the enzyme activity is regulated by a cofactor or an effector, real-time signaling allows facile monitoring of the presence of the regulating molecule. Second, fluorescence signaling creates new opportunities for exploring nucleic acid enzymes for innovative applications, some of which are discussed below.

Many of the reported studies that involve signaling ribozymes and deoxyribozymes share two general features. First, all of the signaling systems are based on the concepts of FDQ

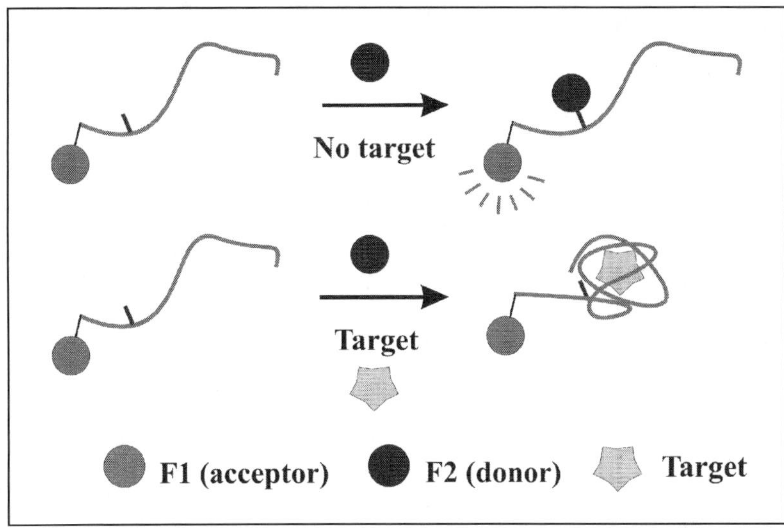

Figure 12. Signaling aptamers by in situ labeling. Target-aptamer complex formation prevents the attachment of the donor fluorophore to an aptamer sequence labeled with the acceptor fluorophore, resulting in the lack of FRET.

(fluorescence de-quenching) or FRET. Second, all of the nucleic acid enzymes employed to date are RNA-cleaving ribozymes or deoxyribozymes. Both FDQ and FRET entail the use of two chromophores that must be brought together for fluorescence quenching or FRET; alternatively, the chromophores must be separated for FDQ or disappearance of FRET. For the latter purpose, a cleavage-based system is a logical choice. The dual labels can be attached onto the catalytic system in two ways: both labels can be attached to the substrate at opposite sides of the cleavage site, or one label can be placed on the substrate and the other on the enzyme at a location that is close to the first label. In both cases, the cleavage will lead to the separation of the two labels and either FDQ or the disappearance of FRET.

FRET and FDQ were initially applied for real-time monitoring of the kinetics of naturally occurring RNA-cleaving ribozymes, such as the hammerhead and hairpin ribozymes. These studies were performed either in vitro[58-61] or in vivo.[62] Jenne et al extended this approach to search for inhibitors of hammerhead ribozyme activity, using a modified substrate containing a fluorophore at one end and a quencher at the other (Fig. 13).[63] When the hammerhead ribozyme is active, cleavage of the substrate separates the fluorophore from the quencher, which leads to an increase in fluorescence intensity via FDQ. However, if the ribozyme is inhibited, the solution will have constant fluorescence intensity. A library of 96 small molecules was tested, and 16 compounds were found to modulate the activity of the hammerhead ribozyme considerably, with K_i varying from 0.1 to 100 μM.

Signaling ribozymes and deoxyribozymes have also been designed as sensors for metal ions, nucleic acids, and proteins. In principle, any metal-ion-dependent ribozyme or deoxyribozyme can be used as a selective metal ion sensor if it requires a particular divalent metal ion as the cofactor for function and is significantly less active in the presence of other metal ions. Li et al explored this principle and successfully designed a lead (Pb^{2+}) sensor using an RNA-cleaving deoxyribozyme named 17E.[64] 17E is a sequence variant of a small DNA enzyme known as 8-17 that has been independently isolated several times by in vitro selection.[29,65-67] 17E was found to exhibit a strong catalytic activity in the presence of Pb^{2+} but significantly reduced activity when

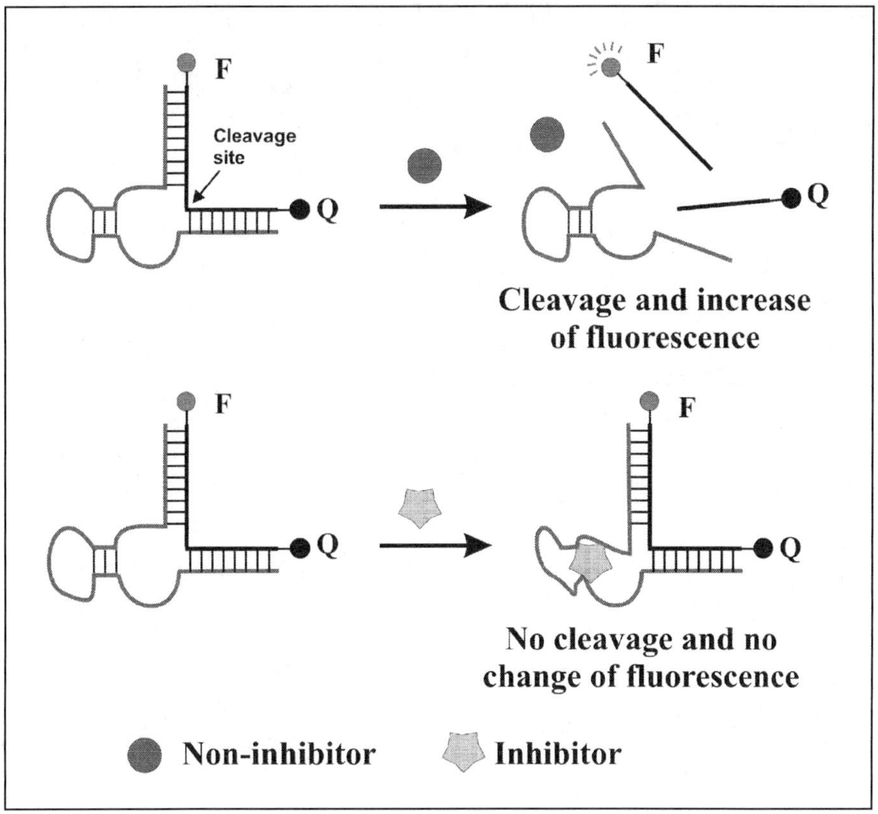

Figure 13. Screening ribozyme inhibitors using signaling ribozymes. The binding of an inhibitor to a ribozyme shuts off both the catalytic activity of the ribozyme and its signaling capability.

other divalent metal ions were supplied. By labeling the substrate strand with a fluorophore (F) and the deoxyribozyme with a quencher (Q), 17E can conveniently report the presence of Pb^{2+} in solution (Fig. 14). Prior to the enzymatic action, the substrate binds tightly to the deoxyribozyme, bringing the fluorophore and the quencher into close proximity for efficient fluorescence quenching. Upon Pb^{2+}-promoted enzymatic action, the deoxyribozyme cleaves the single ribonucleotide linkage embedded in an otherwise all-DNA substrate. The two separated substrate fragments no longer anneal tightly to the deoxyribozyme, resulting in the increase of fluorescence. The catalytic system was ~80-fold more responsive to Pb^{2+} than other divalent metal ions, with a detection range of 10 nM to 4 μM Pb^{2+}.[64,68]

The specificity of deoxyribozyme-based metal ion biosensors can be enhanced by in vitro selection with the incorporation of a negative selection strategy.[69] To eliminate the sequences that are active in the presence of nontarget metal ions, the DNA library is first exposed to a complex "soup" of metal ions that lacks the desired target ion. Next, the sequences that are active in the presence of the metal ion of interest are isolated. Exposing the selected DNA populations to decreased concentrations of the desired metal ion during the later stages of in vitro selection can also increase the affinity of the enzyme for the metal ion cofactor.

In addition to the fluorogenic form of the 17E deoxyribozyme for metal ion sensing, Liu and Lu have also created an elegant colorimetric lead biosensor that can be monitored

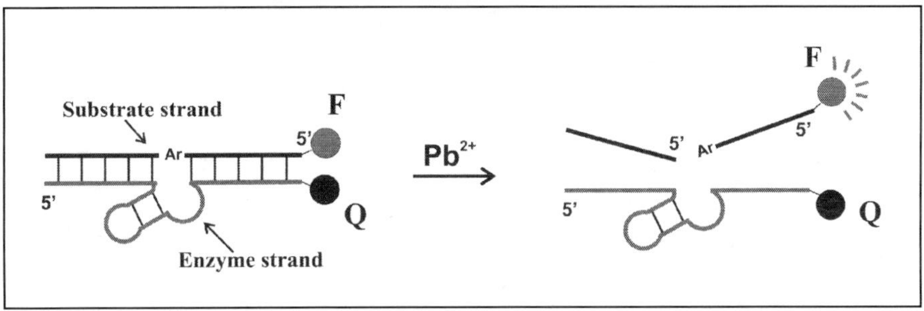

Figure 14. A deoxyribozyme-based Pb^{2+} sensor. The sensor is constructed using an RNA-cleaving deoxyribozyme. Upon addition of Pb^{2+} ions, the deoxyribozyme is activated and cleaves the single ribonucleotide linkage (Ar, adenosine ribonucleotide) in the substrate strand. The cleaved substrate fragments then dissociate from the enzyme, separating the fluorophore from the quencher to generate an increase in fluorescence intensity. This figure is adapted from reference 87.

visually.[68] Unreacted enzyme-substrate complexes mediate the assembly of gold nanoparticles into large colored aggregates. Upon enzyme activation, the assembly of gold nanoparticles dissociates with a concurrent change in color. The sensitivity of the system is tunable to a detection range of 10-200 µM Pb^{2+} through introduction of inactive enzyme strands into the gold nanoparticle network.

The intrinsic ability of nucleic acids to form duplex structures has been exploited in many nucleic acid detection systems, such as the molecular beacon designs described above. Two elegant fluorescence-signaling systems—catalytic molecular beacons[70] and DzyNA-PCR[71]— have recently been developed that ingeniously incorporate the catalytic activities of DNA for nucleic acid detection. Catalytic molecular beacons (CMBs) combine the concepts and advantages of both molecular beacons and deoxyribozymes (Fig. 15).[70] In the first reported example, the previously selected RNA-cleaving deoxyribozyme named 12E was reengineered to contain three intricately linked elements: a hairpin-shaped structural motif (labeled as "Molecular beacon module" in Fig. 15), a DNA enzyme module, and a separate dual-labeled substrate that competes with the MB module for deoxyribozyme binding. In the absence of the target oligonucleotide, the stem of the MB module forms a highly stable hairpin structure with one of the substrate binding arms of the deoxyribozyme, thereby preventing the full interaction of the deoxyribozyme with its substrate. When the target oligonucleotide is added, duplex formation between the target and the MB module disrupts the inhibitory hairpin stem, freeing the previously unavailable substrate-binding arm and restoring enzymatic activity. The substrate is end-labeled with two fluorophores for FRET, and cleavage at the single RNA site abolishes the FRET signal and regenerates the emission signal of the donor fluorophore. The above design strategy is applicable to any RNA-cleaving deoxyribozyme that binds its substrate through duplex formation. One significant advantage of CMBs over standard molecular beacons is their inherent signal-amplification ability, because deoxyribozymes are capable of cleaving substrates with multiple turnover.[70]

In addition to their roles as catalytic platforms for nucleic acid detection, CMBs have been used in molecular computing. Using CMBs as a starting point, Stojanovic et al have made sets of several oligonucleotide-sensitive deoxyribozymes into logic networks able to receive oligonucleotide "inputs" and perform such functions as AND, XOR, and NOT with fluorescent readouts.[72] They also created deoxyribozyme networks able to receive inputs and make decisions, so-called half-adder functions.[73,74] Recently, they used a combination of logic gates made of CMBs regulated by one, two or even three oligonucleotides for the creation of MAYA,

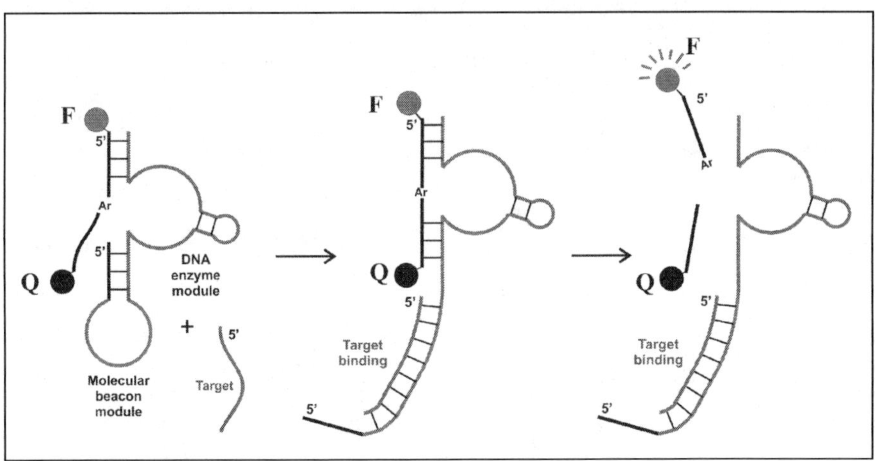

Figure 15. A catalytic molecular beacon. The system uses an MB module to transduce the molecular recognition of an oligonucleotide target to a change in fluorescence intensity, through deoxyribozyme-mediated cleavage of a dual-chromophore labelled substrate. This figure is adapted from reference 87.

an invincible tic-tac-toe DNA player.[74] Although these deoxyribozyme-based computing devices may never rival silicon-based computing, their greatest potential may be in designing "smart" therapeutics that are able both to diagnose and to treat biological problems.[73]

Hartig et al[75-77] have reported several cases of oligonucleotide-regulated ribozymes. For example, they have developed signaling hammerhead ribozymes regulated by an attached RNA aptamer that is responsive to a protein target (Fig. 16).[75] When the protein is absent, the aptamer acts as the antisense oligonucleotide to disrupt substrate binding to the ribozyme, which inhibits its activity. When the protein is present, the aptamer abandons the ribozyme, removing the antisense suppression effect. The activation of the enzyme leads to the cleavage of the fluorogenic substrate and generation of fluorescence. Such aptamer-regulated ribozymes were utilized for the screening of small-molecule inhibitors for a protein target. Three small

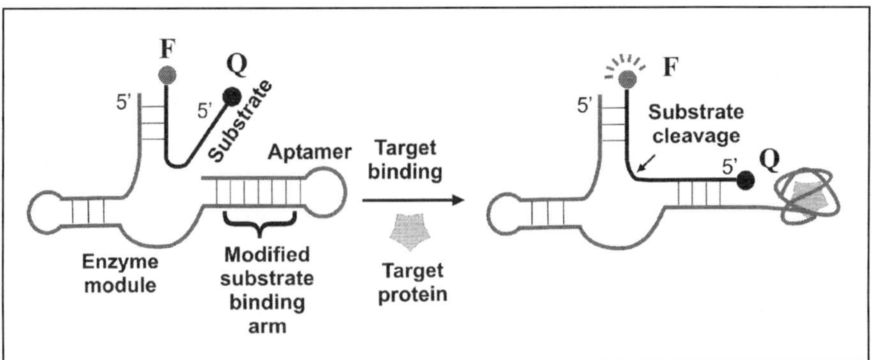

Figure 16. A protein-responsive allosteric signaling ribozyme. The substrate recognition portion of an enzyme is changed to anneal with a portion of the appended aptamer in the absence of the protein target. Binding of the protein to the aptamer disrupts the inhibitory duplex, thereby allowing substrate recognition and catalysis as well as fluorescence signaling. This figure is adapted from reference 88.

molecules that bind HIV-1 REV—coumermycin A_1, nosiheptide and patulin—were identified by this approach.

The invention of DzyNA-PCR, a technique that is capable of monitoring PCR reactions in real time, is another elegant example of exploiting the catalytic activity of DNA for nucleic acid detection (Fig. 17).[71] In this technique, one PCR primer is designed to contain a sequence complementary to the RNA-cleaving 10-23 DNA enzyme. Upon PCR amplification, the corresponding sense strand (containing the 10-23 deoxyribozyme sequence) is generated, which

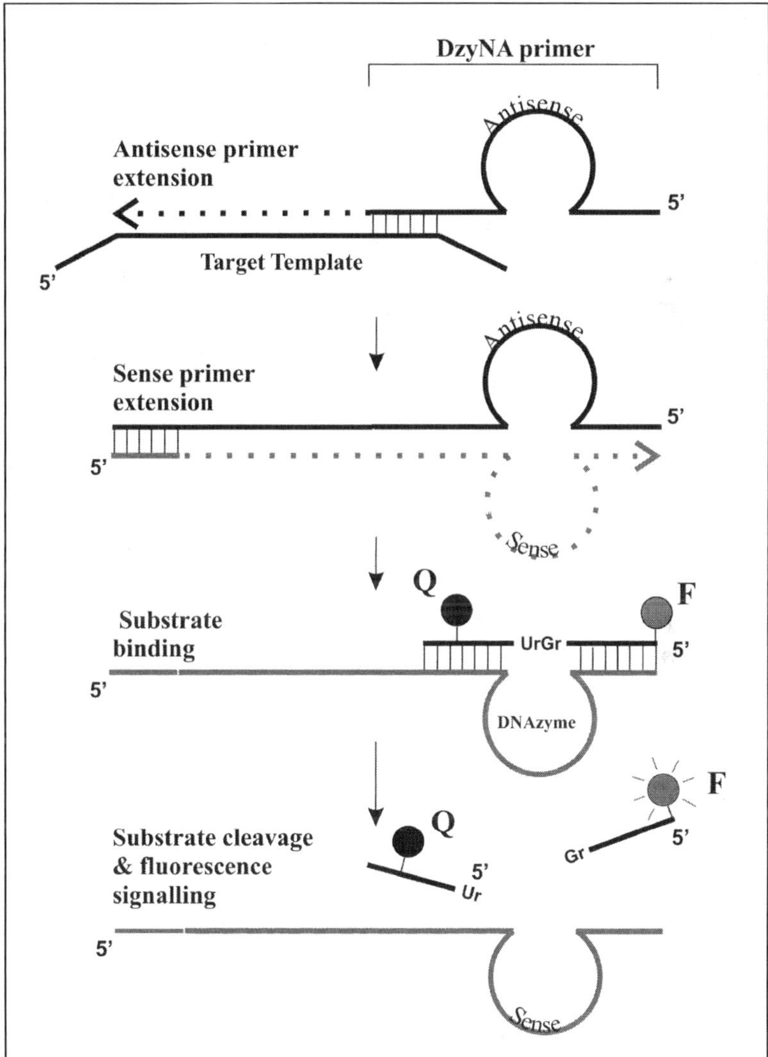

Figure 17. DzyNA-PCR for amplification and detection of specific nucleic acid sequences. The DzyNA primer contains a target-specific sequence in addition to the antisense sequence of an RNA-cleaving deoxyribozyme. During amplification, a deoxyribozyme (as part of the sense strand) is produced that cleaves a fluorophore/quencher-labeled reporter substrate included in the solution. This figure is adapted from reference 87.

can then cleave a substrate modified with a fluorophore-quencher pair. An increase in fluorescence accompanies the amplification of the target DNA as increasing amounts of the active 10-23 DNA enzyme are generated. The DzyNA-PCR system is able to quantitate target DNA (i.e., the PCR template) over six orders of magnitude, from 10^1 to 10^7 molecules.[71] This system also eloquently illustrates the robust nature of deoxyribozymes , because 10-23 can work under buffer conditions and temperatures that are different from those used in the in vitro selection process that first discovered this DNA enzyme.[71]

One significant limitation associated with the end-labeled fluorogenic substrates is high background fluorescence.[70] The inflexibility of modifying preexisting deoxyribozymes limits improvement in this area, because introduction of fluorescent labels closer to the cleavage site usually inactivates deoxyribozymes. Our group has recently conducted two in vitro selection studies to isolate efficient, fluorescence-signaling, RNA-cleaving deoxyribozymes that cleave a chimeric RNA/DNA substrate in which a lone RNA linkage is flanked immediately on either side by a fluorophore and a quencher (Fig. 18A). Prior to the catalytic action of the deoxyribozyme, the fluorophore and quencher are located within a short distance of each other, resulting in very efficient fluorescence quenching. Upon cleavage of the RNA linkage and subsequent product dissociation, the fluorophore moves away from the quencher, leading to intense fluorescence signaling. In the initial study,[78] we created an effective RNase

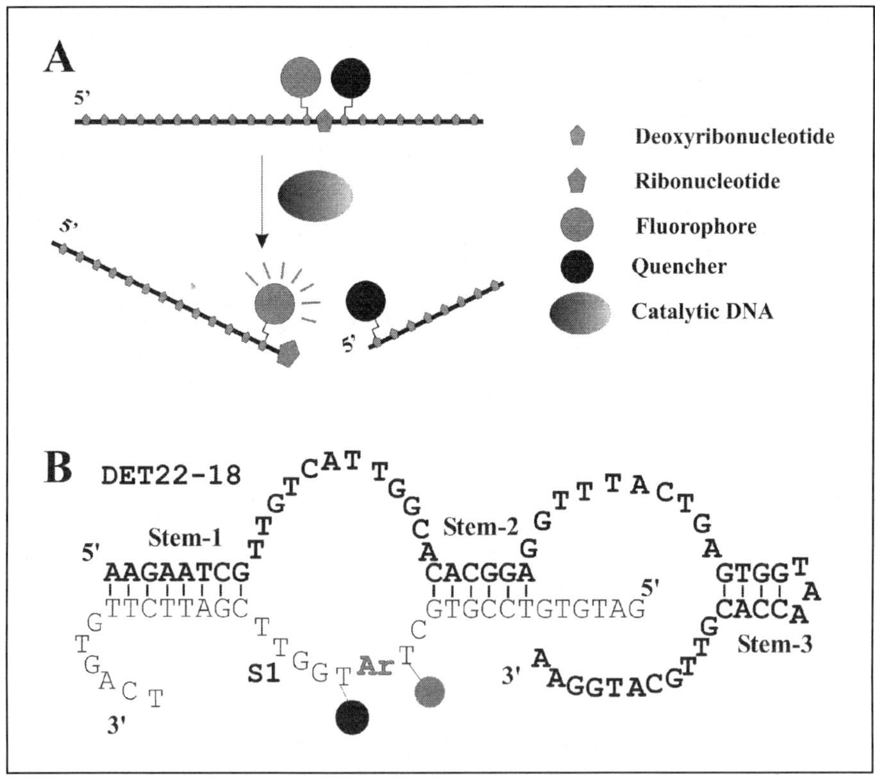

Figure 18. Signaling deoxyribozymes with closely located fluorophore and quencher pairs. A) Catalytic scheme. B) Sequence and secondary structure of a trans-acting signaling deoxyribozyme denoted DET22-18. This figure is adapted from reference 78.

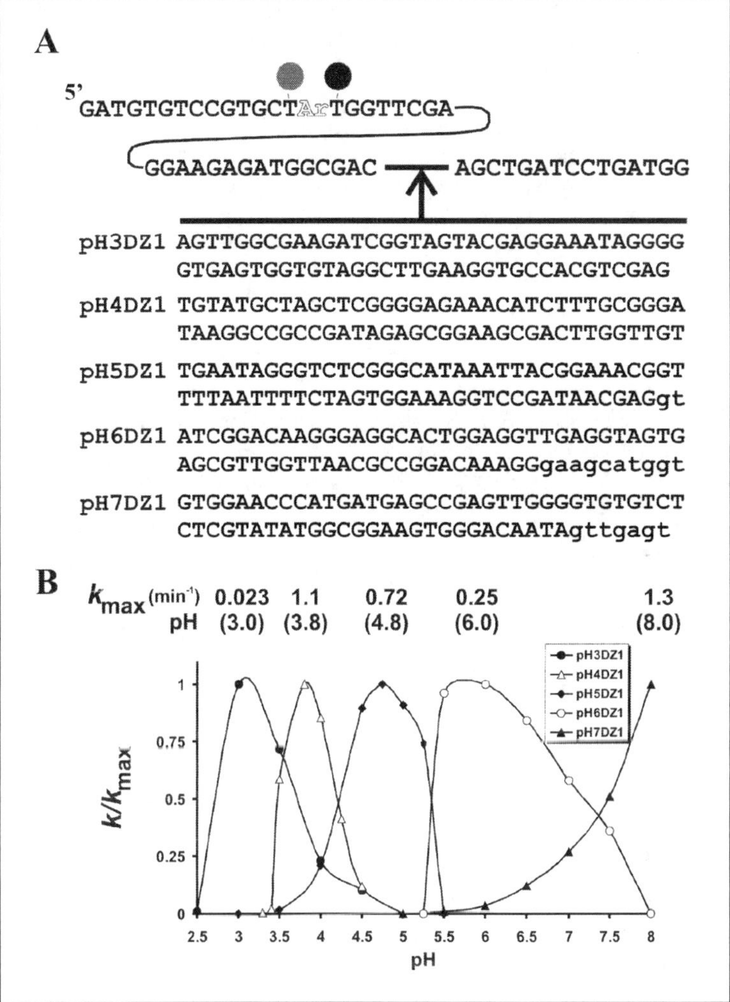

Figure 19. Signaling deoxyribozymes with different pH profiles. A) The sequences of five signaling deoxyribozymes identified by in vitro selection. The lowercase letters represent nucleotides that are not essential for catalysis. B) Normalized catalytic rates versus pH values for the five deoxyribozymes. k is the rate constant at a given pH and k_{max} is the maximum rate observed in each data series. This figure is adapted from reference 79.

denoted DET22-18 (Fig. 18B). DET22-18 can perform multiple turnover catalysis with a k_{cat} of ~7 min⁻¹. Following this study, we created a series of RNA-cleaving signaling deoxyribozymes with various pH optima in the range from pH 3-8 (Fig. 19).[79] The deoxyribozymes were selected first by subjecting the DNA population to multiple rounds of in vitro selection at pH 4. Subsequently, evolution into five different pH-dependent deoxyribozyme pools was achieved by conducting five parallel paths of in vitro evolution at pH 3, 4, 5, 6 and 7. The dominant species from each pH pool were designated as pH3DZ1, pH4DZ1, pH5DZ1, pH6DZ1, and pH7DZ1 (Fig. 19A). Interestingly, the four DNA enzymes selected at pH 3-6 displayed maximum catalytic rates at or near their selection pH (Fig. 19B). The only catalytic DNA for which no pH optimum was detected was pH7DZ1, whose catalytic rate constant increased from pH 5.5 to 8.0. Most of the DNA enzymes exhibited relatively large catalytic rate constants with k_{obs} values ranging between 0.2 and 1.3 min^{-1}; only pH3DZ1 was significantly less efficient, with a k_{obs} of 0.023 min^{-1}. All five DNA enzymes were evaluated for their ability to produce a fluorescence enhancement upon catalysis. The existence of a series of signaling deoxyribozymes covering 5 pH units broadens the utility of signaling DNA enzymes for practical biosensing applications. Further engineering of these catalytic DNA species into target-reporting probes is currently underway.

Perspective

The abundant examples presented above illustrate diverse strategies for designing nucleic acid switches and sensors that conveniently transduce an event of molecular recognition and/or catalysis into an easily detectable fluorescence signal. Most studies conducted to date have aimed to demonstrate proof of concept and used artificial targets in "clean" solutions. The next and more challenging step is to adapt these concepts and probes to more realistic applications. For example, one can imagine the construction of a multiplexed system for simultaneously tracking key components of a living cell (mRNAs, proteins, and metabolites) under a specific set of conditions or under the influence of external stimuli. This would allow the study of cellular processes on a more comprehensive scale, offering more of a global view rather than the limited information acquired by one approach or one technology at a time. Although the application of fluorescent nucleic acid probes for studying living systems is still at an incipient stage, the limited results obtained so far are exciting and promising. For example, Matsuo and colleagues have conducted a study aimed to visualize mRNA in living cells using MBs. They designed a MB targeting the mRNA for basic fibroblast growth factor (bFGF) and successfully visualized the mRNA of bFGF in a human cell line.[80] Similarly, MBs have been successfully applied to detect vav protooncogene mRNA in K562 human leukemia cells[81] and β-actin mRNA in single living kangaroo rat kidney cells.[82] Recently, Bratu et al has conducted a study to follow the transport and localization of mRNA in *Drosophila* oocytes.[37] By microinjecting a nuclease-resistant MB for a target sequence of the *oskar* mRNA and following the hybridization kinetics in real time, mRNA transport was visualized from its birthplace to its final destination. To date, there are no research papers that describe the use of signaling aptamers or signaling nucleic acid enzymes for in vivo imaging of small molecules or proteins. However, aptamers have been shown to function inside cells.[83] Therefore, we anticipate that ongoing efforts will lead to the creation of signaling aptamers and signaling nucleic acid enzymes for in vivo applications.

Perhaps one of the most valuable uses of signaling aptamers and signaling nucleic acid enzymes is in the field of drug discovery. As discussed in this chapter, signaling aptamers and allosteric nucleic acid enzymes can be engineered to report both protein enzymes and small-molecule metabolites. Therefore, these nucleic acid probes may be extremely useful as fluorescent reporters in screening assays for drug discovery, and their potential is currently under investigation in several academic and industrial labs. Two different concepts have been developed to date. The first approach is based on an affinity competition for a particular protein between a nucleic acid enzyme and a small molecule. For example, if the protein binds an allosteric signaling ribozyme, the resulting catalytic activity generates a fluorescent signal. In contrast, if the protein prefers to bind to a small molecule such as a potential drug, then the signaling ribozyme remains inactive, and no fluorescence can be detected.[75] The second approach is designed to detect the activity of an enzyme by monitoring the consumption of the reactants or formation of the products. Consider a reaction

A→B mediated by an enzyme E. If a signaling aptamer has different affinities for A and B, then the enzymatic transformation of A to B is transduced into a change in fluorescence intensity. Consequently, compounds that inhibit E can be identified by monitoring the lack of fluorescence change.[84] The same approach can be used for the case of metabolite-responsive allosteric DNA or RNA enzymes.[85] Although the reported examples are currently at the proof-of-concept stage, the encouraging results hold promise for extensive use of signaling aptamers and signaling nucleic acid enzymes in high-throughput drug discovery in the years to come.

References

1. Updike SJ, Hicks GP. The enzyme electrode. Nature 1967; 214:986-988.
2. Updike SJ, Hicks GP. Reagentless substrate analysis with immobilized enzymes. Science 1967; 158:270-272.
3. Nakamura H, Karube I. Current research activity in biosensors. Anal Bioanal Chem 2003; 377:446-468.
4. Subrahmanyam S, Piletsky SA, Turner AP. Application of natural receptors in sensors and assays. Anal Chem 2002; 74:3942-3951.
5. D'Orazio P. Biosensors in clinical chemistry. Clin Chim Acta 2003; 334:41-69.
6. Pancrazio JJ, Whelan JP, Borkholder DA et al. Development and application of cell-based biosensors. Ann Biomed Eng 1999; 27:697-711.
7. Rizzuto R, Pinton P, Brini M et al. Mitochondria as biosensors of calcium microdomains. Cell Calcium 1999; 26:193-199.
8. Wijesuriya DC, Rechnitz GA. Biosensors based on plant and animal tissues. Biosens Bioelectron 1993; 8:155-160.
9. Ye L, Haupt K. Molecularly imprinted polymers as antibody and receptor mimics for assays, sensors and drug discovery. Anal Bioanal Chem 2004; 378:1887-1897.
10. Southern EM. Detection of specific sequences among DNA fragments separated by gel electrophoresis. J Mol Biol 1975; 98:503-517.
11. Tuerk C, Gold L. Systematic evolution of ligands by exponential enrichment: RNA ligands to bacteriophage T4 DNA polymerase. Science 1990; 249:505-510.
12. Ellington AD, Szostak JW. In vitro selection of RNA molecules that bind specific ligands. Nature 1990; 346:818-822.
13. Famulok M. Oligonucleotide aptamers that recognize small molecules. Curr Opin Struct Biol 1999; 9:324-329.
14. Wilson DS, Szostak JW. In vitro selection of functional nucleic acids. Annu Rev Biochem 1999; 68:611-647.
15. Fahlman RP, Sen D. DNA conformational switches as sensitive electronic sensors of analytes. J Am Chem Soc 2002; 124:4610-4616.
16. Fritz J, Baller MK, Lang HP et al. Translating biomolecular recognition into nanomechanics. Science 2000; 288:316-318.
17. Liss M, Petersen B, Wolf H et al. An aptamer-based quartz crystal protein biosensor. Anal Chem 2002; 74:4488-4495.
18. Potyrailo RA, Conrad RC, Ellington AD et al. Adapting selected nucleic acid ligands (aptamers) to biosensors. Anal Chem 1998; 70:3419-3425.
19. Lakowicz JR. Principles of fluorescence spectroscopy. 2nd ed. New York: Kluwer Academic/Plenum, 1999.
20. Arnold MA. Fiber-optic biosensors. J Biotechnol 1990; 15:219-228.
21. Wolfbeis OS. Fiber-optic chemical sensors and biosensors. Anal Chem 2002; 74:2663-2677.
22. Breaker RR. In vitro selection of catalytic polynucleotides. Chem Rev 1997; 97:371-390.
23. Sen D, Geyer CR. DNA enzymes. Curr Opin Chem Biol 1998; 2:680-687.
24. Li Y, Breaker RR. Deoxyribozymes: New players in the ancient game of biocatalysis. Curr Opin Struct Biol 1999; 9:315-323.
25. Jaschke A. Artificial ribozymes and deoxyribozymes. Curr Opin Struct Biol 2001; 11:321-326.

26. Emilsson GM, Breaker RR. Deoxyribozymes: New activities and new applications. Cell Mol Life Sci 2002; 59:596-607.

27. Carmi N, Shultz LA, Breaker RR. In vitro selection of self-cleaving DNAs. Chem Biol 1996; 3:1039-1046.

28. Breaker RR, Joyce GF. A DNA enzyme that cleaves RNA. Chem Biol 1994; 1:223-229.

29. Santoro SW, Joyce GF. A general purpose RNA-cleaving DNA enzyme. Proc Natl Acad Sci USA 1997; 94:4262-4266.

30. Flynn-Charlebois A, Wang Y, Prior TK et al. Deoxyribozymes with 2'-5' RNA ligase activity. J Am Chem Soc 2003; 125:2444-2454.

31. Li Y, Breaker RR. Phosphorylating DNA with DNA. Proc Natl Acad Sci USA 1999; 96:2746-2751.

32. Breaker RR. Engineered allosteric ribozymes as biosensor components. Curr Opin Biotechnol 2002; 13:31-39.

33. Tyagi S, Kramer FR. Molecular beacons: Probes that fluoresce upon hybridization. Nat Biotechnol 1996; 14:303-308.

34. Marras SA, Kramer FR, Tyagi S. Multiplex detection of single-nucleotide variations using molecular beacons. Genet Anal 1999; 14:151-156.

35. Tan W, Fang X, Li J et al. Molecular beacons: A novel DNA probe for nucleic acid and protein studies. Chem Eur J 2000; 6:1107-1111.

36. Zhang P, Beck T, Tan W. Design of a molecular beacon DNA probe with two fluorophores. Angew Chem Int Ed 2001; 40:402-405.

37. Bratu DP, Cha BJ, Mhlanga MM et al. Visualizing the distribution and transport of mRNAs in living cells. Proc Natl Acad Sci USA 2003; 100:13308-13313.

38. Tyagi S, Marras SA, Kramer FR. Wavelength-shifting molecular beacons. Nat Biotechnol 2000; 18:1191-1196.

39. Nutiu R, Li Y. Tripartite molecular beacons. Nucleic Acids Res 2002; 30:e94.

40. Dubertret B, Calame M, Libchaber AJ. Single-mismatch detection using gold-quenched fluorescent oligonucleotides. Nat Biotechnol 2001; 19:365-370.

41. Du H, Disney MD, Miller BL et al. Hybridization-based unquenching of DNA hairpins on Au surfaces: Prototypical "molecular beacon" biosensors. J Am Chem Soc 2003; 125:4012-4013.

42. Kostrikis LG, Tyagi S, Mhlanga MM et al. Spectral genotyping of human alleles. Science 1998; 279:1228-1229.

43. Bar-Ziv R, Libchaber A. Effects of DNA sequence and structure on binding of RecA to single-stranded DNA. Proc Natl Acad Sci USA 2001; 98:9068-9073.

44. Li JJ, Geyer R, Tan W. Using molecular beacons as a sensitive fluorescence assay for enzymatic cleavage of single-stranded DNA. Nucleic Acids Res 2000; 28:e52.

45. Rizzo J, Gifford LK, Zhang X et al. Chimeric RNA-DNA molecular beacon assay for ribonuclease H activity. Mol Cell Probes 2002; 16:277-283.

46. Liu J, Feldman P, Chung TD. Real-time monitoring in vitro transcription using molecular beacons. Anal Biochem 2002; 300:40-45.

47. Nilsson M, Gullberg M, Dahl F et al. Real-time monitoring of rolling-circle amplification using a modified molecular beacon design. Nucleic Acids Res 2002; 30:e66.

48. Jhaveri S, Kirby R, Conrad R et al. Designed signaling aptamers that transduce molecular recognition to changes in fluorescence intensity. J Am Chem Soc 2000; 122:2469-2473.

49. Yamana K, Ohtani Y, Nakano H et al. Bis-pyrene labeled DNA aptamer as an intelligent fluorescent biosensor. Bioorg Med Chem Lett 2003; 13:3429-3431.

50. Jhaveri S, Rajendran M, Ellington AD. In vitro selection of signaling aptamers. Nat Biotechnol 2000; 18:1293-1297.

51. Hamaguchi N, Ellington A, Stanton M. Aptamer beacons for the direct detection of proteins. Anal Biochem 2001; 294:126-131.

52. Yamamoto R, Baba T, Kumar PK. Molecular beacon aptamer fluoresces in the presence of Tat protein of HIV-1. Genes Cells 2000; 5:389-396.

53. Stojanovic MN, de Prada P, Landry DW. Aptamer-based folding fluorescent sensor for cocaine. J Am Chem Soc 2001; 123:4928-4931.

54. Li JJ, Fang X, Tan W. Molecular aptamer beacons for real-time protein recognition. Biochem Biophys Res Commun 2002; 292:31-40.

55. Fang X, Sen A, Vicens M et al. Synthetic DNA aptamers to detect protein molecular variants in a high-throughput fluorescence quenching assay. ChemBioChem 2003; 4:829-834.
56. Nutiu R, Li Y. Structure-switching signaling aptamers. J Am Chem Soc 2003; 125:4771-4778.
57. Merino EJ, Weeks KM. Fluorogenic resolution of ligand binding by a nucleic acid aptamer. J Am Chem Soc 2003; 125:12370-12371.
58. Perkins TA, Wolf DE, Goodchild J. Fluorescence resonance energy transfer analysis of ribozyme kinetics reveals the mode of action of a facilitator oligonucleotide. Biochemistry 1996; 35:16370-16377.
59. Walter NG, Burke JM. Real-time monitoring of hairpin ribozyme kinetics through base-specific quenching of fluorescein-labeled substrates. RNA 1997; 3:392-404.
60. Jenne A, Gmelin W, Raffler N et al. Real-time charazterization of ribozymes by fluorescence resonance energy transfer (FRET). Angew Chem Int Ed 1999; 9:1300-1303.
61. Singh KK, Parwaresch R, Krupp G. Rapid kinetic characterization of hammerhead ribozymes by real-time monitoring of fluorescence resonance energy transfer (FRET). RNA 1999; 5:1348-1356.
62. Vitiello D, Pecchia DB, Burke JM. Intracellular ribozyme-catalyzed trans-cleavage of RNA monitored by fluorescence resonance energy transfer. RNA 2000; 6:628-637.
63. Jenne A, Hartig JS, Piganeau N et al. Rapid identification and characterization of hammerhead-ribozyme inhibitors using fluorescence-based technology. Nat Biotechnol 2001; 19:56-61.
64. Li J, Lu Y. A highly sensitive and selective catalytic DNA biosensor for lead Ions. J Am Chem Soc 2000; 122:10466-10467.
65. Faulhammer D, Famulok M. The Ca(II) ion as a cofactor for a novel RNA-cleaving deoxyribozyme. Angew Chem Int Ed 1996; 35:2809-2813.
66. Li J, Zheng W, Kwon AH et al. In vitro selection and characterization of a highly efficient Zn(II)-dependent RNA-cleaving deoxyribozyme. Nucleic Acids Res 2000; 28:481-488.
67. Cruz RPG, Withers JW, Li Y. Dinucleotide junction cleavage versatility of 8-17 deoxyribozyme. Chem Biol 2004; 11:57-67.
68. Liu J, Lu Y. A colorimetric lead biosensor using DNAzyme-directed assembly of gold nanoparticles. J Am Chem Soc 2003; 125:6642-6643.
69. Bruesehoff PJ, Li J, Augustine AJ et al. Improving metal ion specificity during in vitro selection of catalytic DNA. Comb Chem High Throughput Screen 2002; 5:327-335.
70. Stojanovic MN, de Prada P, Landry DW. Catalytic molecular beacons. ChemBioChem 2001; 2:411-415.
71. Todd AV, Fuery CJ, Impey HL et al. DzyNA-PCR: Use of DNAzymes to detect and quantify nucleic acid sequences in a real-time fluorescent format. Clin Chem 2000; 46:625-630.
72. Stojanovic MN, Mitchell TE, Stefanovic D. Deoxyribozyme-based logic gates. J Am Chem Soc 2002; 124:3555-3561.
73. Stojanovic MN, Stefanovic D. Deoxyribozyme-based half-adder. J Am Chem Soc 2003; 125:6673-6676.
74. Stojanovic MN, Stefanovic D. A deoxyribozyme-based molecular automaton. Nat Biotechnol 2003; 21:1069-1074.
75. Hartig JS, Najafi-Shoushtari SH, Grune I et al. Protein-dependent ribozymes report molecular interactions in real time. Nat Biotechnol 2002; 20:717-722.
76. Hartig JS, Famulok M. Reporter ribozymes for real-time analysis of domain-specific interactions in biomolecules: HIV-1 reverse transcriptase and the primer-template complex. Angew Chem Int Ed 2002; 41:4263-4266.
77. Hartig JS, Grune I, Najafi-Shoushtari SH et al. Sequence-specific detection of MicroRNAs by signal-amplifying ribozymes. J Am Chem Soc 2004; 126:722-723.
78. Mei SH, Liu Z, Brennan JD et al. An efficient RNA-cleaving DNA enzyme that synchronizes catalysis with fluorescence signaling. J Am Chem Soc 2003; 125:412-420.
79. Liu Z, Mei SH, Brennan JD et al. Assemblage of signaling DNA enzymes with intriguing metal-ion specificities and pH dependences. J Am Chem Soc 2003; 125:7539-7545.
80. Matsuo T. In situ visualization of messenger RNA for basic fibroblast growth factor in living cells. Biochim Biophys Acta 1998; 1379:178-184.
81. Sokol DL, Zhang X, Lu P et al. Real time detection of DNA. RNA hybridization in living cells. Proc Natl Acad Sci USA 1998; 95:11538-11543.

82. Perlette J, Tan W. Real-time monitoring of intracellular mRNA hybridization inside single living cells. Anal Chem 2001; 73:5544-5550.

83. Famulok M, Verma S. In vivo-applied functional RNAs as tools in proteomics and genomics research. Trends Biotechnol 2002; 20:462-466.

84. Nutiu R, Yu JMY, Li Y. An aptamer-based fluorescence assay for monitoring enzymatic activity and enzyme inhibitor screening. ChemBioChem 2004; 5:1139-1144.

85. Srinivasan J, Cload ST, Hamaguchi N et al. ADP-specific sensors enable universal assay of protein kinase activity. Chem Biol 2004; 11:499-508.

86. Nutiu R, Li Y. Structure-switching signaling aptamers: Transducing molecular recognition into fluorescence signaling. Chem Eur J 2004; 10:1868-1876.

87. Schlosser K, Mei SHJ, Li Y. Catalytic DNAs as reporter molecules. Recent Research Developments in Chemistry. India, Trivandrum: Research Signpost, 2003:1.

88. Achenbach JC, Nutiu R, Li Y. Structure-switching allosteric deoxyribozymes. Anal Chim Acta 2005; 534:41-51.

SECTION II
Natural Nucleic Acid Switches and Sensors

CHAPTER 5

Protein-Induced RNA Switches in Nature

Oliver Mayer, Nikolai Windbichler, Herbert Wank and Renée Schroeder*

Abstract

The conformational flexibility of RNA is the basis for its functional versatility. RNA molecules can fold into functionally diverse structures, providing the grounds for great regulatory potential. RNA function can be turned on or off in very short time scales merely through, for example, the binding of ligands. Conformational switching of RNA molecules can be induced by diverse signals, ranging from large molecules like proteins and RNAs to small molecules like coenzymes, amino acids, or nucleotides to temperature and the translation speed of the ribosome. In this chapter, we discuss switches for RNA folding and function induced by naturally occurring RNA-protein interactions.

The conformational flexibility of RNA is also the basis for the RNA folding problem. The numerous structures that an RNA molecule encounters on its folding pathway represent kinetic traps, which can slow down the folding process significantly. In nature, a large number of proteins with different modes of action help RNA molecules to switch conformations. These proteins can stabilize specific structures or nonspecifically resolve low-stability conformers. The latter activity has been called RNA chaperone activity, a function that has been attributed to a large and ever-growing number of proteins. Specialized RNA helicases hydrolyze ATP to resolve stable structures. Noncoding RNAs, which induce functional switches by base pairing, have difficulties identifying their targets due to the generally low accessibility of supposedly unpaired bases. The *E. coli* protein Hfq helps noncoding RNAs to solve this problem. There are many ways by which proteins participate in and regulate RNA structure and as a consequence its function. The diversity of these mechanisms is just beginning to be understood and will be discussed in this chapter.

Introduction

RNA molecules are linear polyanionic chains that must fold into defined three-dimensional structures to accomplish tasks as diverse as catalysis, ligand binding, and protein recognition. The folding pathway of RNA molecules is hierarchical, starting with the formation of short double-stranded helices via Watson-Crick (WC) type base pairs, which are part of the secondary structure. These short secondary structure elements then organize themselves in space to form very diverse tertiary structures.[1] Apart from the canonical AU and GC pairs, RNA supports GU pairs, so the number of possible pairing combinations is very high and the alternative conformations are often comparably stable. This structural versatility is a challenge for an RNA

*Corresponding Author: Renée Schroeder—Max F. Perutz Laboratories, University Departments at the Vienna Biocenter, Department of Microbiology and Genetics, Dr. Bohrgasse 9/4, A-1030, Vienna, Austria. Email: renee.schroeder@univie.ac.at

Nucleic Acid Switches and Sensors, edited by Scott K. Silverman. ©2006 Landes Bioscience and Springer Science+Business Media.

molecule, because it becomes more difficult to reach its native folding state.[2] On the other hand, this structural promiscuity opens the way to an important attribute: the possibility for one RNA molecule to exist in different conformational states that have different functions. An extreme example of an RNA sequence that folds into two functionally completely different states is a designed variant of an in vitro selected RNA ligase that can also fold into a hepatitis delta virus ribozyme. This RNA sequence has two identities, that of an artificial RNA ligase and that of an HDV ribozyme.[3]

The notion that RNAs have flexible conformations arose with the discovery of attenuation at the tryptophan operon by the pioneering work of Charles Yanofski.[4] Several decades later, the phenomenon of structural variation of messenger RNA induced by small ligands and leading to different functions was discovered. These naturally occurring structural rearrangements induced by small molecules have been termed "riboswitches" (see Chapter 6). In the past few years, an unexpectedly large number of riboswitches have been discovered. Ron Breaker's laboratory in particular published a series of papers describing RNA conformational switches induced by small metabolites.[5-7] Binding of these small molecules to the 5'-untranslated region of an mRNA leads to conformational changes resulting in translation inhibition, transcription attenuation, or RNA cleavage. In general, the signals that can induce a switch in RNA conformation and function are diverse, ranging from large molecules like proteins or tRNAs, small molecules like cobalamin, FMN, guanosine or lysine to temperature, the speed of transcription or of the translating ribosome.[8-10] Small, noncoding RNAs have also been reported to induce switches in RNA conformation to regulate mRNA stability and translation.[11]

By summarizing and comparing the latest experiments, this chapter discusses how proteins can induce conformational switches to regulate the function of RNA molecules. We distinguish between different classes of proteins according to their mode of action. A protein can alter the structure of an RNA molecule in several ways. If the RNA is structurally unstable, leading to a mixed population of conformers, binding of a specific protein can stabilize one conformation, thus favoring it over the others. Tight binding of a protein to one part of the RNA might not be enough to unfold stable alternative conformations, requiring the unwinding activity of RNA helicases. In such a case, exemplified by the *Neurospora crassa* tyrosyl tRNA synthetase that recognizes group I intron RNAs, the specific RNA-binding protein recruits a specific RNA helicase, which consumes ATP to unfold the nonnative structure.[12] Another type of protein activity is attributed to Hfq, which binds RNA molecules less specifically and induces the unfolding of neighboring domains, making these accessible to additional binding partners. Sm and Sm-like proteins, which are required for the conformational flexibility of large functional RNAs, may share some properties with Hfq. Finally, we discuss a large family of proteins, such as the nucleocapsid peptide NCp7 from HIV and the *E. coli* StpA protein, that have RNA chaperone activity.[13,14] These proteins are diverse in function but have the common property that they resolve RNA structures in a rather nonspecific way.

Proteins That Bind RNA with High Affinity and Stabilize Specific Structures

In nature, most RNA-based reactions take place in conjunction with proteins. Many RNAs are associated with a specific protein factor that binds tightly to a defined structure, stabilizing it and reducing the population of alternative conformations that act as folding traps (Fig. 1). Studying protein-dependent RNA folding is tedious if the folded state of the RNA has to be determined for each step. Therefore, the use of catalytic RNAs has proven to be especially suitable, because the native state correlates with function. Self-splicing introns that depend on protein cofactors are good systems for addressing the question of how a protein binds to a specific RNA target and how it aids in the folding process. For example, the group I intron-specific splicing factor CYT-18 is a dimeric tRNA synthetase, which stabilizes the group I intron core by

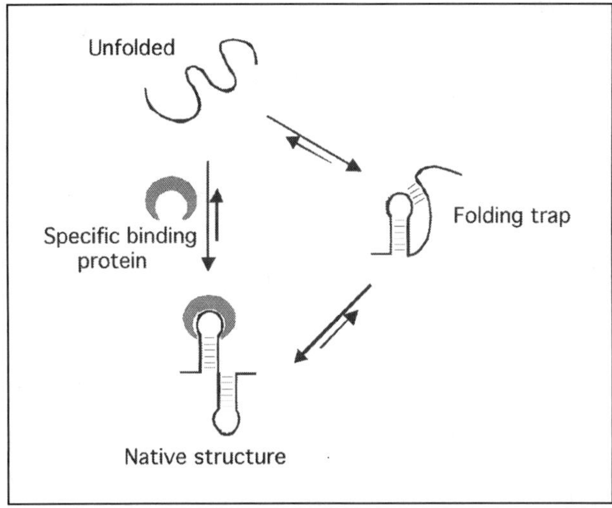

Figure 1. Specific RNA-binding protein. RNA molecules can often form several alternative conformations (folding traps), slowing down the formation of the native structure. Proteins that specifically bind to RNA recognize specific features that are stabilized. At the same time, the population of alternative conformations is reduced.

binding the P4-P6 domain.[15] In contrast, the maturase I-AniI encoded by the *Aspergillus nidulans* mitochondrion binds rapidly to the COB pre-RNA, forming a specific yet labile encounter complex that must be resolved to form a native functional complex in a rate-limiting step. The maturase preassociates with the unfolded RNA and probably facilitates correct folding by resolving misfolded RNAs or by reducing the number of possible conformations; after the intron is correctly folded, the protein locks the RNA in its native state by specific binding.[16] A third example of a specific interaction between a protein and its target group I intron RNA is the yeast CBP2 protein, which facilitates splicing of the mitochondrial bI5 intron. The bI5 intron has been shown to form a collapsed state prior to folding into the native state in the presence of 7 mM Mg^{2+}, whereas it forms an expanded structure at low Mg^{2+} concentrations.[17] The CBP2 protein binds to the intron RNA, which is in its roughly globular native-like collapsed state, and converts it into the native state without mediating global conformational rearrangements, instead promoting only the final few angstroms of RNA folding.[18] Specific RNA-binding proteins have in common that they form tight complexes with their target RNAs, but how these proteins bind and how they guide the RNA to their native states can vary substantially.

RNA Helicases

DexD/H-box proteins are putative RNA helicases that unwind double-stranded RNA (dsRNA) in an energy-dependent fashion using a nucleoside triphosphate, preferentially ATP.[19-22] Two groups of proteins are RNA helicases: the DEAD-box and the DexH-box proteins. Both are named according to their conserved amino acid sequence motifs.[23,24] From the large number of DexD/H-box proteins, only a few have been proven to possess RNA helicase activity.[25] Putative RNA helicases are present in almost all organisms ranging from bacteria and viruses to humans and are involved in a large variety of cellular processes, including nuclear transcription, pre-mRNA splicing, ribosome biogenesis, nucleocytoplasmic transport, translation, RNA decay, and organellar gene expression.[26] The best-studied DEAD-box protein is the eukaryotic initiation factor eIF4A.[27,28] Some of these proteins may even be involved in disrupting or

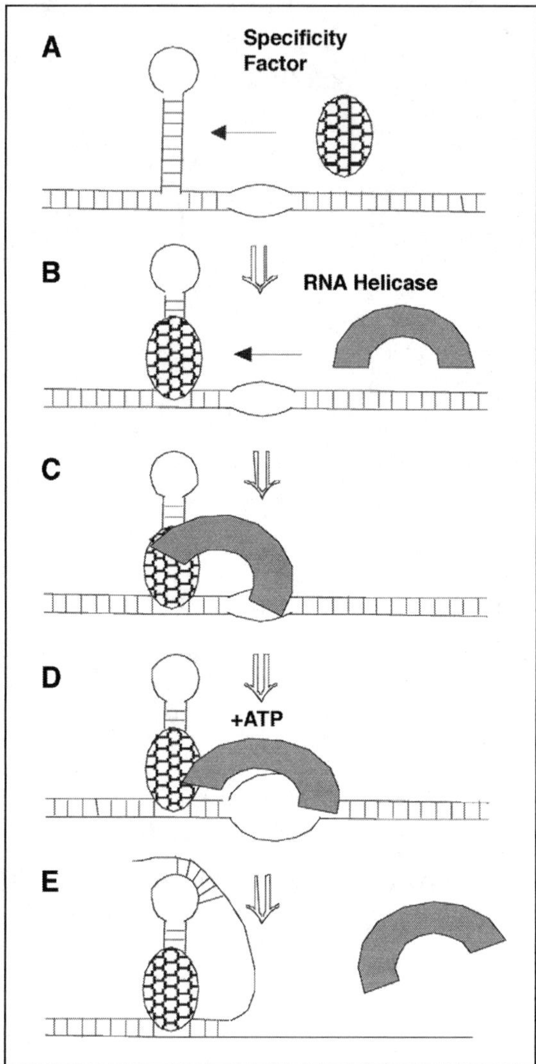

Figure 2. Model for RNA helicase activity. The specificity factor (honeycombed) binds to the target sequence and/or structure (A). The RNA helicase (shaded gray) recognizes the specificity factor (B), thus forming a ternary complex (C). Upon hydrolysis of ATP, the helicase opens helical structures (D). These now-unpaired strands are free to form new base-pairing interactions (E).

rearranging RNA-protein interactions.[21,25,29] In contrast to DNA helicases, the RNA counterparts are thought to modulate formation of only short RNA duplexes in a one-step reaction.

There are similarities in the tertiary structures of DNA and RNA helicases.[30] In vitro, RNA helicases show only limited substrate specificity, whereas in vivo, specificity is necessary in order to prevent random opening of RNA structures and RNA-protein interactions. Therefore, RNA helicases need a specificity factor to identify their place of action (Fig. 2). From the very large number of RNA helicases, we focus here on two interesting exemplary studies: the active disruption of an RNA-protein interaction by the viral DexH/D RNA helicase NPH-II, and the DEAD-box protein CYT-19 from *Neurospora crassa* that uses CYT-18 as specificity factor for finding its substrate.[12,25]

NPH-II

The first in vitro demonstration of the active disruption of an RNA-protein interaction was shown with the vaccinia virus NPH-II DexH/D helicase.[25] The disruption of an RNA-protein interaction has been named RNPase or RNP displacement activity. In this report, the well-characterized U1A protein and its natural target binding site, the 3'-UTR of the U1A mRNA, were used.[31] U1A binds to this target as a dimer, interacting with two asymmetric loops. In order to fulfill its function as an RNA helicase substrate, the RNA was altered by removing a hairpin loop and extending the flanking helical regions. A 3'-overhang was also included to provide a platform for NPH-II binding. This altered substrate still binds U1A very tightly. Using this system, a basic kinetic mechanism of helicase action was established. The dissociation rate for U1A was increased by several orders of magnitude when NPH-II and ATP were present. Although NPH-II appears to "stumble" when it encounters the U1A protein, it maintains its processivity.[21] Thus, NPH-II is able to displace U1A and continue unwinding the RNA substrate, without falling off during the course of reaction.

CYT-19

CYT-19 is a DEAD-box protein isolated from *Neurospora crassa* using a cold-sensitive group I intron splicing-deficient mutant.[12,32] As expected for a DEAD-box protein, CYT-19 is an RNA-dependent ATPase. The purified protein does not specifically bind to the group I intron RNA but requires the group I intron splicing factor CYT-18 as specificity factor. Thus, this RNA helicase functions only in concert with CYT-18 to promote group I intron splicing. The addition of CYT-19 and ATP increases both the rate of splicing and the total yield of spliced RNA, suggesting that CYT-19 resolves misfolded RNAs. Using the well-characterized folding pathway of the *Tetrahymena* LSU group I intron, it was shown that CYT-19 acts by disrupting a nonnative secondary structure that acts as a kinetic trap in the folding of this intron. Binding of CYT-18 to the intron RNA already greatly improves the stability of the RNA, but the reaction does not go to completion, with a significant percentage of the population remaining trapped in misfolded conformations. Recruitment of the CYT-19 RNA helicase leads to resolution of these kinetic traps, and folding proceeds to completion.

Hfq, A Protein That Assists RNA Molecules to Anneal

RNA antisense approaches in research or in therapy are often hampered by the low accessibility of target RNA regions due to local secondary structure, even in supposedly unstructured mRNAs. This problem is also encountered by some *E. coli* noncoding RNAs (ncRNAs), which are a class of naturally occurring antisense RNAs that function in post-transcriptional regulation. Recently, genome-wide screens have brought the number of small RNAs in *E. coli* to around 50, and a large number of these RNAs may regulate post-transcriptional gene expression.[11] These RNAs fulfill their function by partial base pairing to target mRNA transcripts, thereby blocking ribosome access to the Shine-Dalgarno sequence; by opening up inhibitory mRNA structures; or by influencing mRNA stability. The conformational switches in the 5'-UTR of mRNAs in these cases are induced by ncRNAs.

NcRNAs and their targets often show discrete structures in regions that later must anneal with the interaction partner. Whereas the structure of ncRNAs and their target RNAs may have evolved to facilitate their mutual interaction, it is believed that protein factors enhance these RNA-RNA interactions. A candidate for this activity in *E. coli* is the protein Hfq, which was first described as a cofactor required for replication of the RNA phage Qβ.[33] Hfq is strikingly conserved in a wide range of bacteria and highly abundant.[34] It has been estimated that there are approximately 30,000 to 60,000 Hfq molecules per bacterial cell.[35] Hfq has been shown to interact with, or to be required for the activity of, several small interfering RNAs, including DsrA,[36] RprA,[37] OxyS,[38,39] Spot42[34] and RyhB.[40]

The Hfq-dependent action of the small RNAs RyhB and OxyS is particularly enlightening with regard to the mechanism of Hfq activity on these RNAs (Fig. 3). RyhB regulates superoxide dismutase B (sodB) mRNA translation and stability. SODs eliminate free superoxide radicals and are therefore key enzymes of the cellular defense system against oxidative stress. The RyhB-sodB mRNA interaction requires the presence of Hfq and occurs over a stretch of nine complementary nucleotides, encompassing the AUG initiation codon of translation on the sodB mRNA. RyhB binding blocks the translation initiation codon of sodB and triggers the RNase E dependent degradation of both RyhB and sodB mRNA.[41] In this system, it is the mRNA and not the small regulatory RNA that is bound by Hfq with high affinity.[42] Hfq binds to sodB mRNA with a dissociation constant of $K_d = 1.8$ nM. Hfq binds to RyhB much less strongly with a $K_d = 1.5$ μM, which is the same order of magnitude as that for binding of Hfq to DsrA, OxyS or Spot42 RNA.[34,36,39,43]

Because the affinity of Hfq for RyhB RNA is almost three orders of magnitude lower than for sodB mRNA, it is interesting to look at these interactions at the structural level. The Hfq binding site on sodB mRNA is an A/U-rich single-stranded linker of 14 nucleotides between two stem-loops; the first stem-loop (stem-loop a) begins with the first transcribed nucleotide (Fig. 3). The second stem-loop (stem-loop b) encompasses the region in which translation is initiated. Following addition of Hfq, the A/U-rich region is protected against RNase E cleavage, indicating that Hfq binds to and protects this region. In strong contrast, addition of Hfq changes the structure of the stem-loop b region, rendering it more accessible to RyhB. Since this region is complementary to the RyhB RNA, Hfq facilitates binding of RyhB to sodB mRNA by resolving a base-pairing interaction that interferes with binding of RyhB.

In a similar way, Hfq mediates interaction of fhlA mRNA and the small RNA OxyS. OxyS RNA is a 109-nucleotide untranslated RNA that is induced in response to oxidative stress in *E. coli* and activates or represses multiple genes.[44] Among the repressed genes are rpoS and the transcriptional activator fhlA. OxyS repression of fhlA is achieved through two base-pairing interactions, one site that overlaps with the ribosome-binding site and the second site that resides in the fhlA mRNA coding sequence.[45,46] OxyS binding prevents ribosome access to the fhlA mRNA and leads to the repression of translation. Hfq was shown to increase OxyS interaction with its target mRNAs by formation of supershifted complexes that did not depend on the continued presence of Hfq, which was digested away by proteinase K.[39] OxyS RNA consists of three stem-loops and a single-stranded A/U-rich region. While some positions became more protected upon Hfq binding, mainly in the single-stranded region downstream of stem-loop b, some positions outside the binding site showed increased accessibility. Thus, Hfq seems to function by increasing the accessibility of an RNA sequence that is required for interaction with a target RNA.

Interestingly, Hfq shows some characteristic features of eukaryotic Sm proteins, which consist of a group of seven proteins that form a heteroheptameric ring structure and bind to U-rich sequences of the spliceosomal U RNAs. Hfq has in common with the Sm proteins the presence of the Sm1 motif, formation of an oligomeric ring structure, and preferential binding to poly-U RNA. Crystal structures of the *Staphylococcus aureus* Hfq and of a complex of Hfq bound to the oligoribonucleotide AUUUUG provide some insight into the molecular details of Hfq action.[47] Hfq forms a symmetric hexameric ring with a diameter of ~65 Å and a width of ~23 Å (Fig. 3B). The hexamer has a doughnut shape with a central hole of ~12 Å diameter. The Hfq-RNA structure reveals that the single-stranded 7-mer oligoribonucleotide binds in a

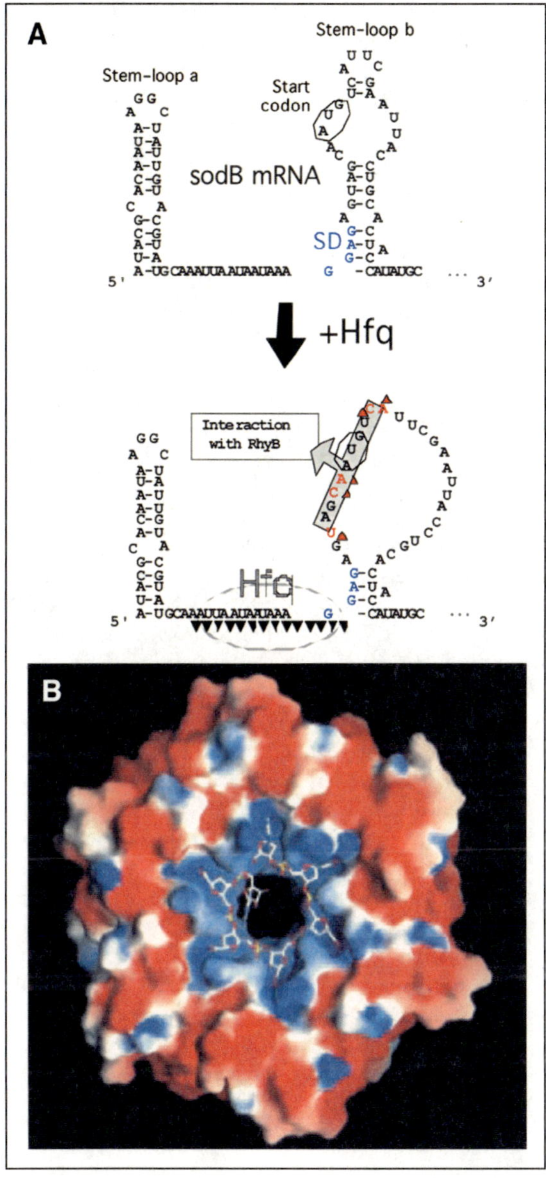

Figure 3. Hfq promotes RNA-RNA interactions. A) The 5'-UTR of the sodB mRNA is shown with stem-loops a and b. The Shine-Dalgarno sequence is marked in blue, the start condon is circled, and the binding site for Hfq is indicated. The structural changes in the sodB mRNA 5'-UTR upon binding of Hfq are shown. RNase footprinting experiments with different RNases were done in the presence and absence of Hfq.[38] Red triangles represent RNase cleavages that are enhanced after binding of Hfq. Black triangles are RNase cleavage sites that are protected upon the addition of Hfq. The sequence complementary to RyhB RNA is shown within a gray square. B) Electrostatic surface representation of the RNA binding site of the Hfq hexamer. Blue is electropositive and red is electronegative. The RNA is shown as a stick model, with oxygen, nitrogen, carbon and phosphorus atoms colored red, blue, white and yellow, respectively. Figure adapted from reference 47.

circularly unwound conformation around the central basic cleft in the pore of one face of the Hfq hexamer. The structural data explain the preferential binding of Hfq to A/U-rich RNA and also provide insight into the mode of action of Hfq on RNA-RNA interactions: when Hfq binds single-stranded RNA, the RNA is unwound within its central pore. The structure indicates that Hfq could accommodate A/U-rich RNA for up to six nucleotides. Such binding and unwinding strongly destabilizes surrounding RNA structures that are located several nucleotides on either side of the binding site, thereby permitting new RNA-RNA interactions.

Proteins with RNA Chaperone Activity

RNA molecules may fold into nonnative but stable structures that lie along the folding pathways to the native conformation.[32,48-51] These alternative structures represent kinetic traps; they have long lifetimes due to the energy required to break the incorrect interactions. RNA duplexes have a very high thermodynamic stability. An RNA helix of 10 base pairs (bp) in length has a dissociation half-life of about 30 min; G/C-rich 10-bp duplexes might have dissociation halftimes of up to 100 years at 30°C.[52,53] Alternative folding states already exist in relatively small RNA molecules like tRNAs. For tRNA^{Leu} it has been shown that an inactive conformation is stable on the hours timescale and that it can be converted to an active structure upon heating in the presence of Mg^{2+}.[54]

How can RNA overcome this folding problem and reach its active, native conformation? One possible solution is the existence of proteins that help folding. Nonspecific RNA-binding proteins solve the kinetic folding problem by acting as RNA chaperones. RNA chaperones are defined as proteins that prevent RNA misfolding and/or resolve misfolded RNAs, thus enabling correct folding of the RNA molecule.[53] The idea of proteins that aid RNA in the folding process was suggested over 20 years ago, when it was shown that a fragment of hnRNP A1 protein can renature kinetically trapped 5S and tRNAs.[55,56] RNA chaperones are a highly diverse family of nucleic acid-binding proteins that exhibit a wide variety of biological activities. No common structural signature has been identified, because they belong to different classes of RNA-binding proteins. In many cases, no known RNA binding domain can be predicted.[57] RNA chaperones seem to recognize neither a common sequence nor a common structural motif.[58-62]

An increasing number of proteins have been shown to act as RNA chaperones. In vitro, RNA chaperone activity can be monitored in different ways. Assays include the measurement of RNA annealing activity, strand exchange activity, and ribozyme turnover stimulation, as well as the promotion of trans-splicing of the T4 *td* intron.[13,14,63-65] Taking advantage of a misfolded species of the T4 phage *td* pre-mRNA in the absence of translation, an in vivo assay for RNA chaperone activity was established. Using this assay, it was possible to show that a series of proteins with in vitro RNA chaperone activity can also resolve a misfolded RNA structure in vivo. The most prominent of these RNA chaperones is the *E. coli* protein StpA.[66-68]

E. coli Protein StpA

StpA was initially isolated as a multicopy suppressor of a splicing-defective mutant of the bacteriophage T4 *td* group I intron.[69] StpA is a 15.3-kDa nucleoid-associated protein present in *Escherichia*, *Shigella* and *Salmonella* species and shares 58% sequence identity with its *E. coli* paralogue H-NS.[69,70] StpA has a nucleic acid binding domain in its C-terminal region, whereas its N-terminus functions as dimerization domain.[71] It was shown that StpA enhances splicing in vitro by using a trans-splicing system, where the *td* intron is split into two half-molecules. RNA strand-annealing reactions can be greatly enhanced by adding StpA. By means of strand-exchange assays, it could be shown that StpA also enhances RNA dissociation, a property required to resolve misfolded structures. StpA is dispensable after assembly of the trans-splicing precursors, because removing the protein with proteinase K treatment before initiating catalysis does not affect the level of splicing enhancement. This demonstrates that StpA is not needed for catalysis, but only for the assembly of the catalytically active group I intron conformation.[14] StpA also induces strand transfer during primer extension by the HIV reverse transcriptase.[72]

In vivo, StpA rescues folding of the *td* pre-mRNA in mutants that are kinetically trapped in nonactive conformations due to the absence of translation.[66,67] In vivo DMS modification of the T4 *td* group I intron RNA showed that in presence of StpA, tertiary structure elements become more accessible to the modifying agent, suggesting that the protein has a destabilizing effect on the overall tertiary structure of the intron. This suggests that the RNA chaperone activity of StpA results partly from the resolution of tertiary interactions in the intron structure. This activity may be a general characteristic of proteins with RNA chaperone activity, but this still remains to be analyzed.[73]

Nucleocapsid Protein of HIV-1

Retroviral nucleocapsid (NC) proteins are examples of viral proteins that exert RNA chaperone activities for several RNA switches during viral replication and assembly. HIV-1 NC is a highly basic protein consisting of 55 amino acids and two zinc finger structures. The nucleic acid chaperone activity of NC greatly accelerates $tRNA^{Lys}$ annealing to the primer-binding site, and it facilitates minus- and plus-strand transfer reactions of retroviral reverse transcriptase.[74,75] The zinc fingers are not essential for $tRNA^{Lys}$ annealing, but they are responsible for inducing subtle tertiary structure changes and for destabilizing helical regions of the tRNA.[76] In contrast, the zinc finger structures are critical for the annealing step in minus-strand transfer.[77]

The HIV-1 genome is packaged as a dimer into a viral particle. The two copies of the RNA are linked through base pairing of the dimer linkage sites (DLS) located in the 5'-leader of the RNA. Studies on the overall structure of the full length HIV-1 leader RNA have revealed the existence of a compactly folded structure of the RNA induced by base-pairing interactions between sequences that would otherwise form the poly-A and the dimerization initiation site (DIS) domains.[78-83] This compact structure masks the DIS loop, and the RNA is in a dimerization-incompetent form. Before the dimerization of the genome, the compact structure is lost and a branched structure containing the poly-A and DIS domains is formed, suggesting that this conformational switch regulates HIV dimerization (Fig. 4). The viral NC protein has been shown to stimulate dimerization of HIV-1 RNA under physiological conditions.[84,85] Using band-shift assays, it could be shown that upon addition of NC protein, the compact structure is disrupted and dimerization is enabled through exposure of the DIS hairpin, which indicates that NC enhances dimerization through stabilization of the branched structure. It favors the branched folding over the thermodynamically more stable rod-like conformer structure of the HIV-1 leader RNA. The dimerization reaction is initiated by base-pairing of the self-complementary DIS loops, forming a kissing complex. NC accelerates the transition of the kissing complex into the more stable extended complex.[86-88] The existence of two different structures of the HIV-1 RNA creates the possibility for the leader RNA to function as a regulatory switch during viral replication. The structural transition initiated by the NC protein may lead to the formation of signals necessary for RNA dimerization, packaging, and also reverse transcription.[89]

The general and nonspecific nature of NC's RNA chaperone activity was further demonstrated through its acceleration of ribozyme activity. NC can facilitate both the formation and disruption of interactions between the hammerhead ribozyme and its substrates and products, thus increasing the rate of catalysis. The protein also resolves kinetically trapped misfolded complexes of the ribozyme.[13,62,64,90]

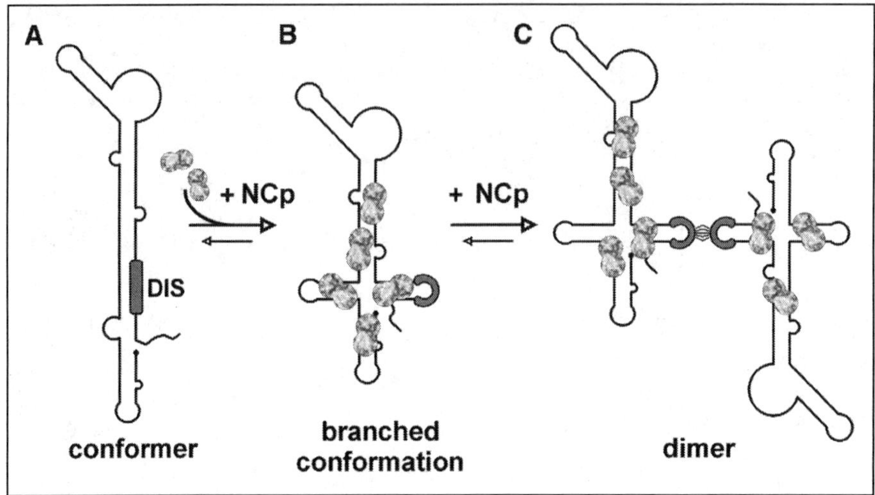

Figure 4. Nucleocapsid protein (NCp) induced rearrangement of the HIV-1 leader RNA structure. In the absence of NCp, the RNA forms the most stable structure (A). Binding of NCp induces a switch of the RNA structure into the branched conformation (B). This rearrangement leads to the formation of the DIS loop (grey), which then dimerizes with another autocomplementary DIS loop via a kissing interaction (C). DIS = dimer initiation signal.[89]

Proteins That Assist with the Formation of RNA-Protein Complexes

One of the most prominent RNA switches known to date is the formation of the U2/U6 snRNP complex from the U4/U6. The spliceosome is highly dynamic, with multiple RNA-RNA and RNA-protein rearrangements occurring during assembly and disassembly. Little is known about how these switches are induced, but we can look at the proteins that build up the snRNPs and those required for its assembly. The spliceosomal snRNPs U1, U2, U4 and U5 contain a common RNP structure termed the Sm core formed by the binding of Sm proteins onto the U snRNAs. The Sm proteins are reminiscent of Hfq in that they form a ring-shaped multimeric structure. In vitro, the Sm proteins bind spontaneously to the snRNAs, but in vivo an additional complex was found to be necessary for the assembly of the spliceosomal core complexes. The SMN complex is named after the "survival of motor neurons" protein implicated in spinal muscular atrophy disease. Reduced expression of SMN results in degeneration of motor neurons and weakness of voluntary muscles.[91] SMN is part of a complex involved in the assembly of the Sm core,[92] which probably facilitates binding of the Sm proteins to the snRNAs. The Sm proteins are modified through symmetrical arginine dimethylation, enhancing the binding of Sm proteins to the SMN complex.

In yeast, no equivalent for the SMN complex has been found. Instead, the La autoantigen might be involved in the facilitation of snRNP assembly. Whether La can functionally replace the SMN protein remains to be tested. The first protein that binds to newly transcribed RNA polymerase III transcripts is the highly conserved La autoantigen. In yeast, this protein, called Lhp1p, is required for tRNA maturation.[93] Another role for La might also be assisting in the assembly of the U6 snRNP. U6 is an RNA polymerase III transcript and in contrast to the other spliceosomal RNAs, U6 is not bound by the Sm proteins, but instead by a set of seven Sm-like proteins, the Lsm proteins.[94] A mutation in the Lsm8p protein results in a reduced level of mature U6 snRNPs, consistent with a defect in U6 snRNP assembly. The exciting fact about this mutant is that it requires the La protein Lhl1p for growth, suggesting that La acts as a chaperone for U6 snRNP assembly.[95] The La protein might resemble Hfq in its mode of action in that it binds a set of RNAs that are relatively structured, thereby facilitating their assembly with other RNAs or proteins. The way in which La chaperones the folding process of its RNA targets is unknown.

Perspective

Our understanding of how proteins modulate RNA structures has improved substantially in the past few years. Experiments have revealed a large variety of interaction modes, both in specificity and in the way that proteins influence RNA folding. The number of well-characterized RNA-protein complexes is increasing steadily, but detailed mechanistic studies have been reported for only some examples. We expect that kinetic and structural analyses of these interactions will reveal as-yet unknown influences of proteins on RNA structure, folding, and function. Studies of dynamic interactions will shed light on the mechanisms involved in the induction of RNA structural changes.

Acknowledgements

We thank all the members of the Schroeder lab for comments and critical reviewing of the manuscript and especially Dr. Paul Watson for the help with English grammar and references. Work in our laboratory is funded by the Austrian Science Fund (FWF) grants F1703, F1704, P16026 and Z72.

References

1. Brion P, Westhof E. Hierarchy and dynamics of RNA folding. Annu Rev Biophys Biomol Struct 1997; 26:113-137.
2. Zuker M. On finding all suboptimal foldings of an RNA molecule. Science 1989; 244:48-52.
3. Schultes EA, Bartel DP. One sequence, two ribozymes: Implications for the emergence of new ribozyme folds. Science 2000; 289:448-452.
4. Yanofsky C. Attenuation in the control of expression of bacterial operons. Nature 1981; 289:751-758.
5. Nahvi A, Sudarsan N, Ebert MS et al. Genetic control by a metabolite binding mRNA. Chem Biol 2002; 9:1043-1049.
6. Winkler W, Nahvi A, Breaker RR. Thiamine derivatives bind messenger RNAs directly to regulate bacterial gene expression. Nature 2002; 419:952-956.
7. Winkler WC, Nahvi A, Roth A et al. Control of gene expression by a natural metabolite-responsive ribozyme. Nature 2004; 428:281-286.
8. Mandal M, Boese B, Barrick JE et al. Riboswitches control fundamental biochemical pathways in Bacillus subtilis and other bacteria. Cell 2003; 113:577-586.
9. Grundy FJ, Henkin TM. tRNA as a positive regulator of transcription antitermination in B. subtilis. Cell 1993; 74:475-482.
10. Henkin TM. Control of transcription termination in prokaryotes. Annu Rev Genet 1996; 30:35-57.
11. Hershberg R, Altuvia S, Margalit H. A survey of small RNA-encoding genes in Escherichia coli. Nucleic Acids Res 2003; 31:1813-1820.
12. Mohr S, Stryker JM, Lambowitz AM. A DEAD-box protein functions as an ATP-dependent RNA chaperone in group I intron splicing. Cell 2002; 109:769-779.
13. Tsuchihashi Z, Khosla M, Herschlag D. Protein enhancement of hammerhead ribozyme catalysis. Science 1993; 262:99-102.
14. Zhang A, Derbyshire V, Salvo JL et al. Escherichia coli protein StpA stimulates self-splicing by promoting RNA assembly in vitro. RNA 1995; 1:783-793.
15. Caprara MG, Lehnert V, Lambowitz AM et al. A tyrosyl-tRNA synthetase recognizes a conserved tRNA-like structural motif in the group I intron catalytic core. Cell 1996; 87:1135-1145.

16. Solem A, Chatterjee P, Caprara MG. A novel mechanism for protein-assisted group I intron splicing. RNA 2002; 8:412-425.
17. Buchmueller KL, Webb AE, Richardson DA et al. A collapsed nonnative RNA folding state. Nat Struct Biol 2000; 7:362-366.
18. Buchmueller KL, Weeks KM. Near native structure in an RNA collapsed state. Biochemistry 2003; 42:13869-13878.
19. de la Cruz J, Kressler D, Linder P. Unwinding RNA in Saccharomyces cerevisiae: DEAD-box proteins and related families. Trends Biochem Sci 1999; 24:192-198.
20. Tanner NK, Linder P. DExD/H box RNA helicases: From generic motors to specific dissociation functions. Mol Cell 2001; 8:251-262.
21. Schwer B. A new twist on RNA helicases: DExH/D box proteins as RNPases. Nat Struct Biol 2001; 8:113-116.
22. Silverman E, Edwalds-Gilbert G, Lin R. DExD/H-box proteins and their partners: Helping RNA hilecases unwind. Gene 2003; 312:1-16.
23. Linder P, Lasko PF, Ashburner M et al. Birth of the D-E-A-D box. Nature 1989; 337:121-122.
24. Gorbalenya AE, Koonin EV. Helicases: Amino-acid sequence comparisons and structure-function relationships. Curr Opin Struct Biol 1993; 3:419-429.
25. Jankowsky E, Gross CH, Shuman S et al. Active disruption of an RNA-protein interaction by a DExH/D RNA helicase. Science 2001; 291:121-125.
26. Rocak S, Linder P. DEAD-box proteins: The driving forces behind RNA metabolism. Nat Rev Mol Cell Biol 2004; 5:232-241.
27. Gingras AC, Raught B, Sonenberg N. eIF4 initiation factors: Effectors of mRNA recruitment to ribosomes and regulators of translation. Annu Rev Biochem 1999; 68:913-963.
28. Rogers Jr GW, Komar AA, Merrick WC. eIF4A: The godfather of the DEAD box helicases. Prog Nucleic Acid Res Mol Biol 2002; 72:307-331.
29. Linder P, Stutz F. mRNA export: Travelling with DEAD box proteins. Curr Biol 2001; 11:R961-963.
30. Korolev S, Yao N, Lohman TM et al. Comparisons between the structures of HCV and Rep helicases reveal structural similarities between SF1 and SF2 super-families of helicases. Protein Sci 1998; 7:605-610.
31. Grainger RJ, Norman DG, Lilley DM. Binding of U1A protein to the 3' untranslated region of its pre-mRNA. J Mol Biol 1999; 288:585-594.
32. Lorsch JR. RNA chaperones exist and DEAD box proteins get a life. Cell 2002; 109:797-800.
33. Franze de Fernandez MT, Eoyang L, August JT. Factor fraction required for the synthesis of bacteriophage Qbeta-RNA. Nature 1968; 219:588-590.
34. Moller T, Franch T, Hojrup P et al. Hfq. A bacterial Sm-like protein that mediates RNA-RNA interaction. Mol Cell 2002; 9:23-30.
35. Kajitani M, Kato A, Wada A et al. Regulation of the Escherichia coli hfq gene encoding the host factor for phage Q beta. J Bacteriol 1994; 176:531-534.
36. Sledjeski DD, Whitman C, Zhang A. Hfq is necessary for regulation by the untranslated RNA DsrA. J Bacteriol 2001; 183:1997-2005.
37. Wassarman KM, Repoila F, Rosenow C et al. Identification of novel small RNAs using comparative genomics and microarrays. Genes Dev 2001; 15:1637-1651.
38. Zhang A, Altuvia S, Tiwari A et al. The OxyS regulatory RNA represses rpoS translation and binds the Hfq (HF-I) protein. EMBO J 1998; 17:6061-6068.
39. Zhang A, Wassarman KM, Ortega J et al. The Sm-like Hfq protein increases Oxys RNA interaction with target mRNAs. Mol Cell 2002; 9:11-22.
40. Masse E, Gottesman S. A small RNA regulates the expression of genes involved in iron metabolism in Escherichia coli. Proc Natl Acad Sci USA 2002; 99:4620-4625.
41. Masse E, Escorcia FE, Gottesman S. Coupled degradation of a small regulatory RNA and its mRNA targets in Escherichia coli. Genes Dev 2003; 17:2374-2383.
42. Geissmann TA, Touati D. Hfq, a new chaperoning role: Binding to messenger RNA determines access for small RNA regulator. EMBO J 2004; 23:396-405.
43. Brescia CC, Mikulecky PJ, Feig AL et al. Identification of the Hfq-binding site on DsrA RNA: Hfq binds without altering DsrA secondary structure. RNA 2003; 9:33-43.
44. Altuvia S, Weinstein-Fischer D, Zhang A et al. A small, stable RNA induced by oxidative stress: Role as a pleiotropic regulator and antimutator. Cell 1997; 90:43-53.

45. Altuvia S, Zhang A, Argaman L et al. The Escherichia coli OxyS regulatory RNA represses fhlA translation by blocking ribosome binding. EMBO J 1998; 17:6069-6075.

46. Argaman L, Altuvia S. fhlA repression by OxyS RNA: Kissing complex formation at two sites results in a stable antisense-target RNA complex. J Mol Biol 2000; 300:1101-1112.

47. Schumacher MA, Pearson RF, Moller T et al. Structures of the pleiotropic translational regulator Hfq and an Hfq-RNA complex: A bacterial Sm-like protein. EMBO J 2002; 21:3546-3556.

48. Herschlag D. RNA chaperones and the RNA folding problem. J Biol Chem 1995; 270:20871-20874.

49. Treiber DK, Williamson JR. Concerted kinetic folding of a multidomain ribozyme with a disrupted loop-receptor interaction. J Mol Biol 2001; 305:11-21.

50. Woodson SA. Recent insights on RNA folding mechanisms from catalytic RNA. Cell Mol Life Sci 2000; 57:796-808.

51. Thirumalai D, Lee N, Woodson SA et al. Early events in RNA folding. Annu Rev Phys Chem 2001; 52:751-762.

52. Turner DH, Sugimoto N, Freier SM. Thermodynamics and kinetics of base-pairing and of DNA and RNA self-assembly and helix coil transition: Springer-Verlag, 1990.

53. Herschlag D. RNA chaperones and the RNA folding problem. J Biol Chem 1995; 270:20871-20874.

54. Lindahl T, Adams A. Native and renatured transfer ribonucleic acid. Science 1966; 152:512-514.

55. Karpel RL, Burchard AC. Physical studies of the interaction of a calf thymus helix-destablizing protein with nucleic acids. Biochemistry 1980; 19:4674-4682.

56. Karpel RL, Miller NS, Fresco JR. Mechanistic studies of ribonucleic acid renaturation by a helix-destabilizing protein. Biochemistry 1982; 21:2102-2108.

57. Cristofari G, Darlix JL. The ubiquitous nature of RNA chaperone proteins. Prog Nucleic Acid Res Mol Biol 2002; 72:223-268.

58. Munroe SH, Dong XF. Heterogeneous nuclear ribonucleoprotein A1 catalyzes RNA-RNA annealing. Proc Natl Acad Sci USA 1992; 89:895-899.

59. Portman DS, Dreyfuss G. RNA annealing activities in HeLa nuclei. EMBO J 1994; 13:213-221.

60. Darlix JL, Lapadat-Tapolsky M, de Rocquigny H et al. First glimpses at structure-function relationships of the nucleocapsid protein of retroviruses. J Mol Biol 1995; 254:523-537.

61. Weeks KM. Protein-facilitated RNA folding. Curr Opin Struct Biol 1997; 7:336-342.

62. Rein A, Henderson LE, Levin JG. Nucleic-acid-chaperone activity of retroviral nucleocapsid proteins: Significance for viral replication. Trends Biochem Sci 1998; 23:297-301.

63. Coetzee T, Herschlag D, Belfort M. Escherichia coli proteins, including ribosomal protein S12, facilitate in vitro splicing of phage T4 introns by acting as RNA chaperones. Genes Dev 1994; 8:1575-1588.

64. Herschlag D, Khosla M, Tsuchihashi Z et al. An RNA chaperone activity of nonspecific RNA binding proteins in hammerhead ribozyme catalysis. EMBO J 1994; 13:2913-2924.

65. Nedbal W, Frey M, Willemann B et al. Mechanistic insights into p53-promoted RNA-RNA annealing. J Mol Biol 1997; 266:677-687.

66. Semrad K, Schroeder R. A ribosomal function is necessary for efficient splicing of the T4 phage thymidylate synthase intron in vivo. Genes Dev 1998; 12:1327-1337.

67. Clodi E, Semrad K, Schroeder R. Assaying RNA chaperone activity in vivo using a novel RNA folding trap. EMBO J 1999; 18:3776-3782.

68. Moll I, Leitsch D, Steinhauser T et al. RNA chaperone activity of the Sm-like Hfq protein. EMBO Rep 2003; 4:284-289.

69. Zhang A, Belfort M. Nucleotide sequence of a newly-identified Escherichia coli gene, stpA, encoding an H-NS-like protein. Nucleic Acids Res 1992; 20:6735.

70. Zhang A, Rimsky S, Reaban ME et al. Escherichia coli protein analogs StpA and H-NS: Regulatory loops, similar and disparate effects on nucleic acid dynamics. EMBO J 1996; 15:1340-1349.

71. Cusick ME, Belfort M. Domain structure and RNA annealing activity of the Escherichia coli regulatory protein StpA. Mol Microbiol 1998; 28:847-857.

72. Negroni M, Buc H. Copy-choice recombination by reverse transcriptases: Reshuffling of genetic markers mediated by RNA chaperones. Proc Natl Acad Sci USA 2000; 97:6385-6390.

73. Waldsich C, Grossberger R, Schroeder R. RNA chaperone StpA loosens interactions of the tertiary structure in the td group I intron in vivo. Genes Dev 2002; 16:2300-2312.

74. De Rocquigny H, Gabus C, Vincent A et al. Viral RNA annealing activities of human immunodeficiency virus type 1 nucleocapsid protein require only peptide domains outside the zinc fingers. Proc Natl Acad Sci USA 1992; 89:6472-6476.

75. Allain B, Lapadat-Tapolsky M, Berlioz C et al. Transactivation of the minus-strand DNA transfer by nucleocapsid protein during reverse transcription of the retroviral genome. EMBO J 1994; 13:973-981.

76. Hargittai MR, Mangla AT, Gorelick RJ et al. HIV-1 nucleocapsid protein zinc finger structures induce tRNA^{Lys,3} structural changes but are not critical for primer/template annealing. J Mol Biol 2001; 312:985-997.

77. Guo J, Wu T, Anderson J et al. Zinc finger structures in the human immunodeficiency virus type 1 nucleocapsid protein facilitate efficient minus- and plus-strand transfer. J Virol 2000; 74:8980-8988.

78. Berkhout B, van Wamel JL. The leader of the HIV-1 RNA genome forms a compactly folded tertiary structure. RNA 2000; 6:282-295.

79. Berkhout B, Klaver B, Das AT. A conserved hairpin structure predicted for the poly(A) signal of human and simian immunodeficiency viruses. Virology 1995; 207:276-281.

80. Laughrea M, Jette L. A 19-nucleotide sequence upstream of the 5' major splice donor is part of the dimerization domain of human immunodeficiency virus 1 genomic RNA. Biochemistry 1994; 33:13464-13474.

81. Skripkin E, Paillart JC, Marquet R et al. Identification of the primary site of the human immunodeficiency virus type 1 RNA dimerization in vitro. Proc Natl Acad Sci USA 1994; 91:4945-4949.

82. Mujeeb A, Clever JL, Billeci TM et al. Structure of the dimer initiation complex of HIV-1 genomic RNA. Nat Struct Biol 1998; 5:432-436.

83. Girard F, Barbault F, Gouyette C et al. Dimer initiation sequence of HIV-1$_{Lai}$ genomic RNA: NMR solution structure of the extended duplex. J Biomol Struct Dyn 1999; 16:1145-1157.

84. Darlix JL, Gabus C, Nugeyre MT et al. Cis elements and trans-acting factors involved in the RNA dimerization of the human immunodeficiency virus HIV-1. J Mol Biol 1990; 216:689-699.

85. Muriaux D, Girard PM, Bonnet-Mathoniere B et al. Dimerization of HIV-1$_{Lai}$ RNA at low ionic strength. An autocomplementary sequence in the 5' leader region is evidenced by an antisense oligonucleotide. J Biol Chem 1995; 270:8209-8216.

86. Laughrea M, Jette L. Kissing-loop model of HIV-1 genome dimerization: HIV-1 RNAs can assume alternative dimeric forms, and all sequences upstream or downstream of hairpin 248-271 are dispensable for dimer formation. Biochemistry 1996; 35:1589-1598.

87. Paillart JC, Marquet R, Skripkin E et al. Dimerization of retroviral genomic RNAs: Structural and functional implications. Biochimie 1996; 78:639-653.

88. Windbichler N, Werner M, Schroeder R. Kissing complex-mediated dimerisation of HIV-1 RNA: Coupling extended duplex formation to ribozyme cleavage. Nucleic Acids Res 2003; 31:6419-6427.

89. Huthoff H, Berkhout B. Two Alternative strcutures of the HIV-1 leader RNA. RNA 2001; 7:143-157.

90. Bertrand EL, Rossi JJ. Facilitation of hammerhead ribozyme catalysis by the nucleocapsid protein of HIV-1 and the heterogeneous nuclear ribonucleoprotein A1. EMBO J 1994; 13:2904-2912.

91. Crawford TO. From enigmatic to problematic: The new molecular genetics of childhood spinal muscular atrophy. Neurology 1996; 46:335-340.

92. Fischer U, Liu Q, Dreyfuss G. The SMN-SIP1 complex has an essential role in spliceosomal snRNP biogenesis. Cell 1997; 90:1023-1029.

93. Wolin SL, Cedervall T. The La protein. Annu Rev Biochem 2002; 71:375-403.

94. Mayes AE, Verdone L, Legrain P et al. Characterization of Sm-like proteins in yeast and their association with U6 snRNA. EMBO J 1999; 18:4321-4331.

95. Pannone BK, Xue D, Wolin SL. A role for the yeast La protein in U6 snRNP assembly: Evidence that the La protein is a molecular chaperone for RNA polymerase III transcripts. EMBO J 1998; 15:7442-74453.

CHAPTER 6

Riboswitches as Genetic Control Elements

Ali Nahvi and Ronald R. Breaker*

Abstract

Riboswitches are metabolite-sensing RNA elements that are present in the noncoding portions of certain messenger RNAs. Each riboswitch carries an aptamer that is highly selective for its target metabolite and an expression platform that more directly interfaces with gene expression systems. In bacteria, changes in RNA structure brought about by ligand binding are harnessed to control expression by modulating mRNA transcription, translation, or RNA stability. The gene control mechanisms and ligand binding characteristics of riboswitches range from minimalist to surprisingly complex. Therefore it is likely that some riboswitches represent the simplest of ways to control genes, while others exhibit levels of sophistication that until now had only been seen with protein genetic factors.

Introduction

Complex living systems must sense a variety of chemical compounds in order to respond to their surroundings and to maintain metabolic stasis. For many years, the role of natural biosensor was thought to be played almost exclusively by protein factors. However, recent studies are revealing the existence of a diversity of metabolite-sensing RNA elements called "riboswitches" (Fig. 1).[1,2] These complex RNA structures are located in specific messenger RNAs where they directly bind to target compounds and control gene expression. Each riboswitch establishes a desired level of expression from an adjoining open reading frame (ORF) by harnessing allosteric changes in RNA folding that are brought about by ligand binding.[1-9]

Ongoing studies are continuing to reveal new classes of riboswitches and to define the scope of this form of gene control system. To date, we know of 11 classes of riboswitches where there is convincing proof that the RNA directly binds the metabolite in the absence of proteins and where evidence for a plausible mechanism for genetic control is available.[10-14] Some of these riboswitch classes are very widespread, whereas others currently appear to be present only in certain bacterial lineages.[6,15] Similarly, some organisms such as *Bacillus subtilis* have approximately 2% of their genes under the control of riboswitches,[10] while some other bacteria so far appear to carry few if any riboswitches.

Interestingly, allosteric RNA molecules had been created by using molecular engineering techniques several years preceding the confirmation that riboswitches exist naturally.[16]

*Corresponding Author: Ronald R. Breaker—Department of Molecular, Cellular and Developmental Biology, Yale University, P.O. Box 208103, New Haven, Connecticut 06520, U.S.A. Email: ronald.breaker@yale.edu

Nucleic Acid Switches and Sensors, edited by Scott K. Silverman. ©2006 Landes Bioscience and Springer Science+Business Media.

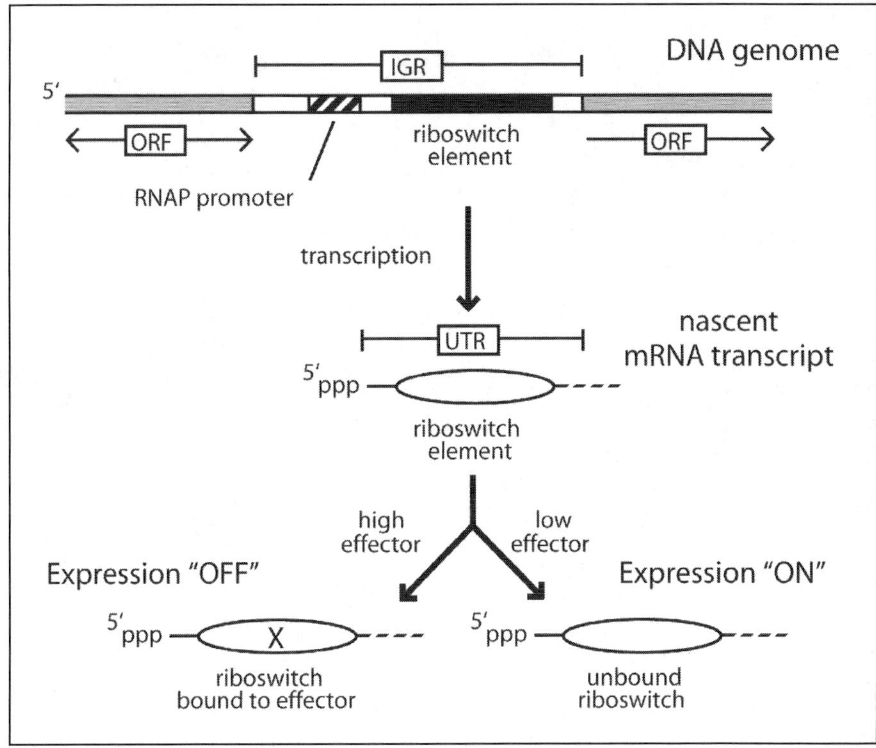

Figure 1. Typical genomic locations of riboswitch elements. Riboswitches usually reside in the intergenic region (IGR) between two open reading frames (ORFs). In many instances, a recognizable promoter sequence for bacterial RNA polymerase (RNAP) occurs within the IGR just upstream of the region of highest sequence conservation (corresponding to the aptamer domain). Initiation of transcription from this promoter in a left-to-right orientation allows the riboswitch to be produced prior to synthesis of the coding portion of the mRNA located immediately downstream. The orientation of the gene upstream of the promoter is not relevant. Most commonly, riboswitches function as genetic "OFF" switches, such that high concentrations of the target metabolite lead to allosteric changes in the 5′ untranslated region (UTR) that cause down-regulation of gene expression.

This chapter provides an overview of the structural and functional features of several classes of natural riboswitches. Similarities between natural and engineered RNAs suggest that additional engineering efforts could be used to create designer riboswitches with tailored ligand specificities and complex mechanisms.

Why RNA Can Serve as a Metabolite-Sensing Genetic Switch

The construction and maintenance of modern cellular life requires that complex biochemical tasks be carried out with exceptional spatial and temporal organization. Proteins, with their 20 common amino acid subunits and extraordinary chemical and structural diversity, are clearly better suited than DNA or RNA to perform many of these tasks. Thus, the discovery in the early 1960s that protein factors can control the expression of genes[17] (and the many subsequent discoveries of similar protein factors) seemed to provide the expected answer for the many genetic control challenges that are faced by cells. Sensing of biochemical signals in most instances will require distinct factors that selectively bind to their corresponding target molecule or otherwise alter their function in response to physical signals such as light and temperature.

However, do all such genetic factors need to be protein? An analogous question once existed in the field of biocatalysis, wherein it was presumed that all enzymes were made of protein. Although the vast majority of chemical transformations in cells are indeed carried out by protein enzymes, there are now nine distinct classes of ribozymes known that catalyze several reactions critical for all cellular life.[11,18,19] If modern cells entrust RNA to form powerful active sites to catalyze some important biochemical processes, then perhaps RNAs might also be used in other instances to form precision biosensors that identify and respond to specific biochemical cues.[20,21]

Mechanisms for genetic control that do not require the direct participation of sensory proteins have been known for some time. For example, transcription attenuation mechanisms are used by some bacteria to activate the expression of genes involved in the biosynthesis of particular amino acids.[22] Here, it is the ribosome that ultimately serves as the sensor of adequate amino acid concentrations. The speed of translation of a leader peptide, whose corresponding coding region precedes the main open reading frame, is reduced only when a specific aminoacyl-tRNA is absent. This slowing or stalling of the ribosome permits the 5′ untranslated region (UTR) to fold into an alternate structure that prevents formation of an intrinsic transcription terminator structure.[23,24] The failure to terminate transcription ultimately promotes gene expression due to increased production of full-length transcript. In contrast, when aminoacyl-tRNAs are present in adequate amounts, rapid translation by the ribosome permits terminator stem formation, leading to transcription termination and down-regulation of gene expression.

In addition, the 5′ UTR of some mRNAs controls gene expression by binding nonaminoacylated tRNAs[25] or by responding to changes in temperature.[26-28] These mechanisms do not necessarily require the direct involvement of protein factors to detect biochemical signals and to actively control gene expression machinery. These more exotic mechanisms hint at a possible larger role for RNA as a direct sensor of metabolic status.

Another line of evidence suggesting that RNA could be used to form metabolite-sensing gene control systems came from molecular engineering experiments. Numerous examples of RNA and DNA aptamers have been generated that fold into complex shapes and selectively bind target molecules.[29-31] Some of these aptamers exhibit binding specificities that are rather striking, thus demonstrating that RNA can achieve levels of binding affinity and discrimination that rival those of many protein factors.[32,33] It is these high levels of binding affinity and specificity that are required for use in a natural setting, where low concentrations of a target metabolite might be present along with high concentrations of closely related biosynthetic intermediates and derivatives.

Identifying the First Riboswitches

Researchers have speculated for some time that RNA might be used as a natural biosensor of metabolic signals. This was fostered in large part by in vitro selection studies that gave rise to so many engineered aptamers. If small ligand-binding aptamers could be isolated from a population of a few billion variant RNAs, then natural aptamers most probably would have emerged from the many diverse RNAs that have been sampled by biology over the last few billion years. This argument was supported by the results of similar RNA engineering experiments that gave rise to allosteric ribozymes. These studies demonstrated that aptamers indeed could have substantial utility if they were fused to functional RNAs in such a way as to induce a change in activity.[34-38]

Another impetus for invoking the existence of metabolite-binding riboswitches was the fact that there were a number of gene control mysteries that could be explained simply by the presence of such an element. For example, it had been known for about a decade that the expression of genes responsible for synthesizing and importing coenzyme B_{12} in certain bacteria was repressed when the levels of this essential coenzyme were adequate.[39-41] Despite several attempts to identify a protein factor that was responsible for sensing coenzyme B_{12} and repressing these genes, no evidence of such a protein was found.[42,43] A similar story existed for riboflavin-mediated control of the expression of genes in certain bacteria that were responsible for biosynthesis of FMN,[44-47] and for yet another involving thiamine pyrophosphate biosynthesis.[48,49] It was these unexplained gene control mysteries that were first examined for possible riboswitch function.

Four publications in 2002 provided the evidence needed to confirm that these gene control mysteries indeed involved the action of metabolite-binding RNAs.[1,2,50,51] A combination of RNA probing and equilibrium dialysis assays confirmed that the most highly conserved portions of the 5′ UTRs of the suspected mRNAs indeed form highly specific receptors that dock with their target metabolites in the complete absence of proteins. In addition, site-directed mutatgenesis experiments and gene expression assays in vivo confirmed that metabolite binding by these RNAs was critical for proper gene control. To date, similar studies that were inspired by published genetic data have confirmed the existence of seven riboswitch classes.[10] More recently, bioinformatics approaches have proven to be useful in identifying new classes of riboswitches, as described below in greater detail.

Dissecting Riboswitches into Functional Components

Riboswitches are typically composed of two functional domains.[52] The first is a natural aptamer (Fig. 2) that selectively binds to its target molecule and establishes the sensing identity of the molecular switch. The second functional domain is called an expression platform (Fig. 3). This serves as an interface between the aptamer and the gene expression machinery, where it transduces metabolite binding into a change in the level of gene expression.

Aptamer domains are used to distinguish each known class of riboswitch because they remain extraordinarily well conserved in sequence and secondary structure, even among distantly related organisms. There are 11 classes of riboswitches confirmed to date (Fig. 2), and these riboswitches uniquely sense a variety of metabolites such as enzyme cofactors, amino acids, nucleotide fragments, and compounds that are only a few transformations removed from entering glycolysis or the citric acid cycle. Most of these target compounds are universal in biology and have remained unchanged through evolution. This might help rationalize why their corresponding RNA aptamers also remain highly conserved. In addition, RNA is made from only four monomers, which severely limits the diversity of changes that can be made to an aptamer without large-scale changes in RNA structure and folding. If the aptamer domains are ancient in origin, then the consensus sequences that we observe today might closely represent the earliest metabolite-sensing RNAs.

In contrast to the highly conserved aptamer domains, expression platforms can vary widely in primary sequence and in mechanism of action (Fig. 3). At least two mechanisms have been proposed to operate in bacterial cells. Evidence from bioinformatics and mutational studies suggest that transcription termination (Fig. 3A) and translation initiation (Fig. 3B) are controlled by metabolite binding via the formation of mutually exclusive base-paired structures. Most commonly, riboswitches serve as genetic "OFF" switches, wherein metabolite binding causes the formation of stem-loop structures that induce transcription termination[10,50,51,53-57] or that occlude ribosome access to the Shine-Dalgarno sequence.[1,2,42,58,59] In the absence of sufficient metabolite concentrations, each expression platform alternatively folds to form a stem-loop structure that competes with the inhibitory stem-loop described above. In rare instances, the reverse mechanism is used, wherein metabolite binding induces gene expression by permitting such anti-inhibitory stems to form preferentially. These genetic "ON" switches are rare only because cells have a greater need to turn off biosynthetic pathways when metabolite concentrations are adequate, compared to their need to activate processes when high concentrations of metabolite are reached.[60]

The notion of distinctive aptamer and expression platform domains might be too simplistic in some cases. When expression platforms have been defined, they typically share nucleotides with the adjacent aptamer domain. For example, it is commonly observed that the initial base-pairing element (P1) of the aptamer participates in the competitive formation of inhibitory and anti-inhibitory structures.[2,10,51,55,57] In a few instances, the expression platform appears to be comprised almost exclusively of nucleotides that form the aptamer domain.[60] In other instances, for example with the *glmS* ribozyme (see below), the precise mechanism by which gene expression is controlled remains undefined. It is possible that metabolite-mediated ribozyme activation might simply promote mRNA degradation.[11] If true, then the notion of a distinctive expression platform domain would not apply.

Simple Riboswitches

The majority of bacterial riboswitches harness the conformational change brought about by the binding of a single ligand to control transcription termination or translation initiation. This one-to-one relationship between ligand and RNA results in a linear response in gene expression to changing concentrations of the molecule being sensed. For example, a 10-fold increase in the concentration of target causes a 10-fold change in the level of gene expression. The complexity of riboswitches that have simple configurations is a function of the complexity of their aptamer domains and the complexity of the more variable expression platforms.

Among the simplest classes of riboswitches known are those that respond to the purine nucleobases guanine and adenine. These riboswitches carry aptamers that have essentially identical consensus sequences and most likely have nearly identical three-dimensional shapes. This similarity is nicely represented by examples of guanine-specific and adenine-specific riboswitches from *B. subtilis*. A guanine riboswitch associated with the *xpt-pbuX* operon (Fig. 4A, top) forms a three-stem junction and carries more than 30 well-conserved nucleotides (Fig. 2E).[10] Likewise, the adenine-specific riboswitch associated with the *ydhL* gene (Fig. 4A, bottom) matches the same consensus sequence and secondary structure.[60] Although the *ydhL* RNA carries 23 mutations relative to the *xpt* RNA, 20 of these mutations occur in putative stem elements, where they largely retain base pairing. Of the remaining three mutations, two occur at nonconserved positions. This leaves only a single C-to-U mutation that resides in an otherwise strictly conserved portion of the element.

The most straightforward explanation for the difference in ligand specificity between the two aptamer variants is that the C or U residue forms a Watson-Crick base pair with its corresponding purine target. The remainder of the architecture of the aptamer presumably remains unchanged, and thus could form identical contacts with parts of the purine ring that are constant between guanine and adenine. This proposal is consistent with the observation that a single nucleotide change at this critical site results in a change in specificity to the other purine for both the *xpt* and *ydhL* RNAs.[60] Recently, atomic-resolution structural models for guanine- and adenine-specific riboswitches have been proposed based on X-ray diffraction data,[61,62] and these structures are consistent with the interpretations described above that were made from biochemical data.

Although both riboswitches make use of transcription termination mechanisms to control gene expression, their expression platforms are sufficiently different such that the

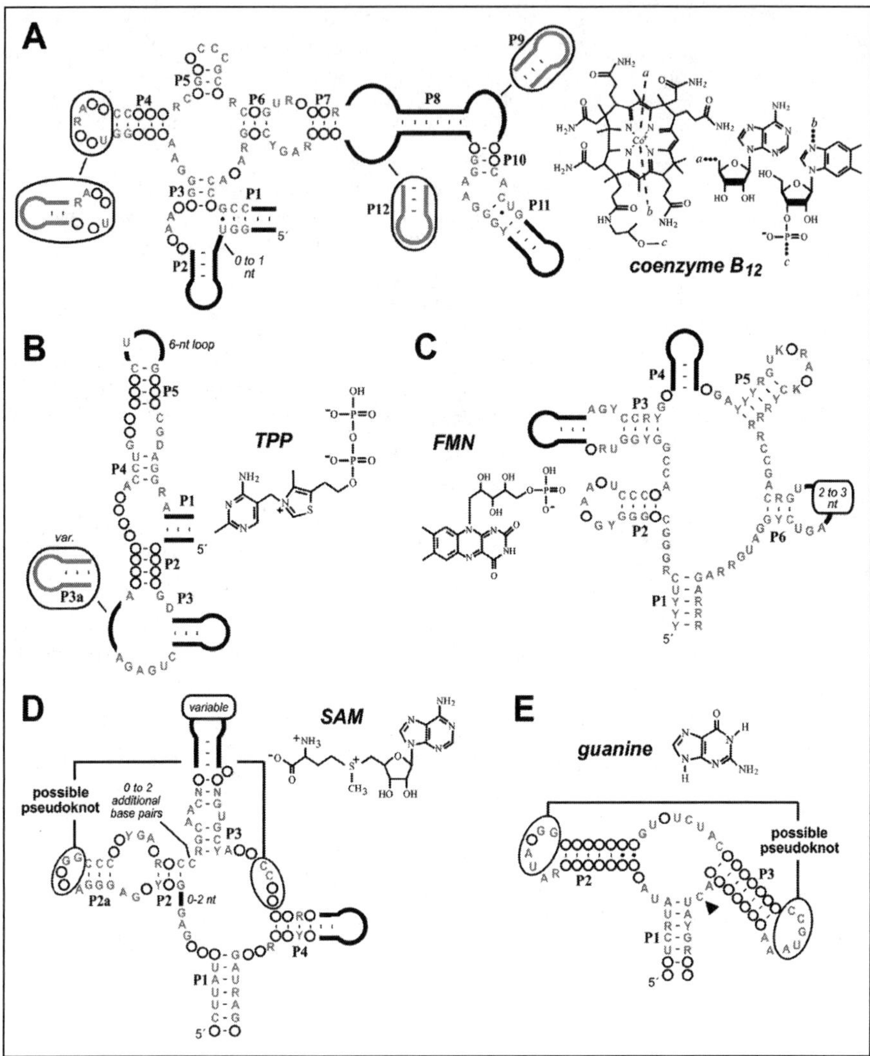

Figure 2. Consensus sequences and secondary structure models for the aptamer domains of known riboswitches. Nucleotides depicted in gray are conserved in greater than 90% of the representatives identified. Circles indicate the presence of a nucleotide whose base identity is not conserved. Letters R and Y represent purine and pyrimidine bases, respectively. In addition, K designates G or U; W designates A or U; H designates A, C or U; D designates G, A, or U; N represents any of the four bases. Additional data for the (A) coenzyme B_{12}, (B) TPP, (C) FMN, (D) SAM I, (E) guanine, (F) adenine, (G) lysine, (H) GlcN6P and (I) glycine riboswitches have been published.[1,2,10-14,51,55,57,60] Details of the (J) SAM II and (K) pre-queuine [pre-Q_1] riboswitches will be published elsewhere. Figure is continued on next page.

xpt RNA is a genetic "OFF" switch, while the *ydhL* RNA is a genetic "ON" switch. The latter adenine riboswitch forms a large intrinsic transcription terminator stem (Fig. 4B, top) in the absence of ligand that prevents RNA polymerase from generating full length

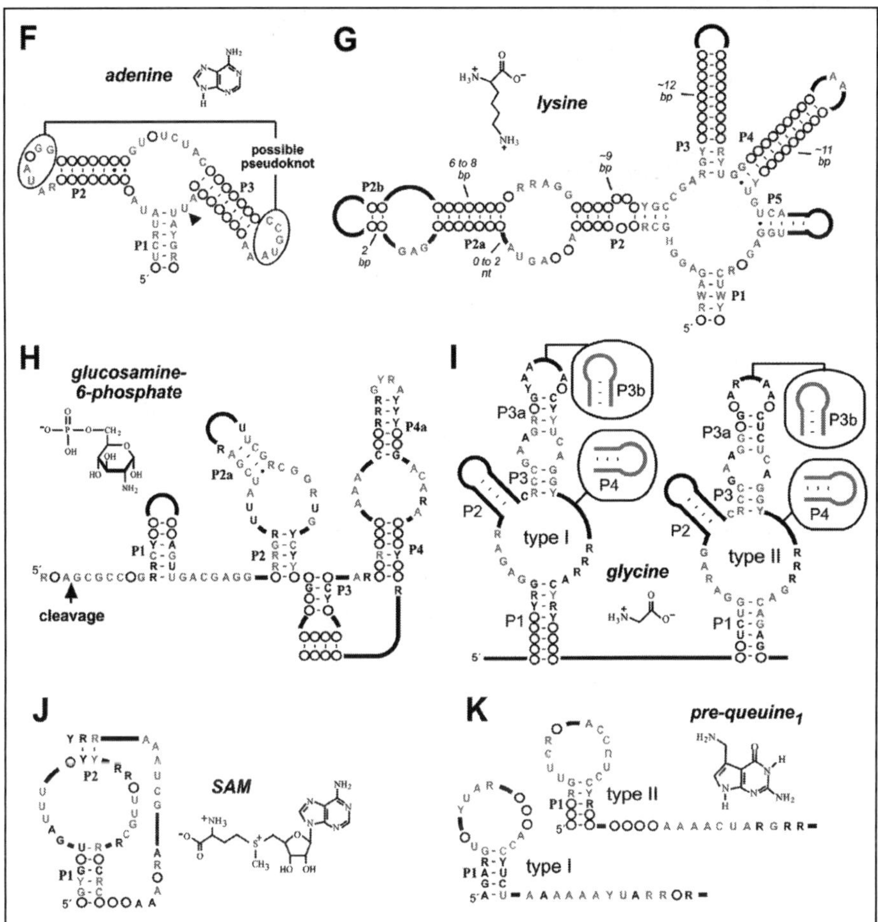

Figure 2, continued.

mRNAs. In contrast, the formation of the secondary structure required for adenine binding (Fig. 4B, bottom), which is expected to be more stable upon docking with adenine, requires the participation of nucleotides that would otherwise be used to form the terminator hairpin.

Similar straightforward mechanisms for alternative secondary structure formation are commonly used by riboswitches. Recent findings indicate that, in some instances, riboswitches become kinetically trapped in such alternately folded structures and do not freely switch between "ON" and "OFF" states on a timescale that is relevant to biology. These "kinetically driven" switches require that a sufficient concentration of metabolite be present at the right time during folding of the nascent 5′ UTR to direct them along a specific folding pathway that leads to the desired genetic outcome.[63] These cases become much more complicated because proper function of the riboswitch depends upon parameters, such as the speed of RNA polymerase as it generates the nascent mRNA, that are separate from the characteristics of the riboswitch RNA when examined in isolation.

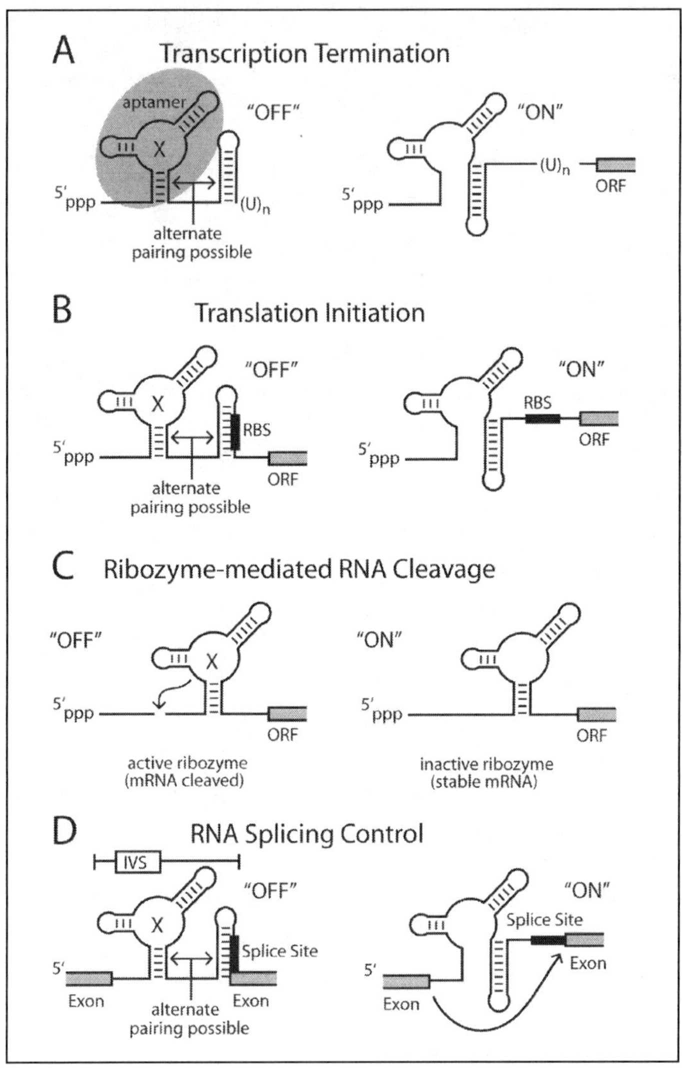

Figure 3. Established and putative gene control mechanisms of riboswitches. A) Riboswitch-mediated control of transcription termination involves the mutually exclusive formation of terminator (left) and anti-terminator (right) hairpins. The intrinsic transcription terminator (hairpin followed by a run of U residues) causes RNA polymerase to terminate transcription before the ORF is reached. B) Riboswitch-mediated control of translation initiation typically involves similar mutually exclusive formation of stem structures. Formation of an anti-ribosome binding site (RBS) structure prevents the mRNA from being bound by ribosomes. Alternatively, formation of an anti-anti-RBS structure liberates the RBS for interaction with ribosomes for subsequent translation. C) A "ribozyme riboswitch" triggers RNA cleavage in a metabolite-dependent fashion. The metabolite could be an allosteric effector, or it could be a coenzyme that directly participates in the catalytic core of the ribozyme. D) Possible mechanism for the riboswitch-mediated control of pre-mRNA splicing. Alternative base pairing established by metabolite binding causes occlusion of nucleotides that are critical for removal of an intervening sequence (IVS), which is required before gene expression is possible. In this instance, blocking of the 3′ intron-exon junction prevents splicing, thus precluding the expression of the interrupted ORF.

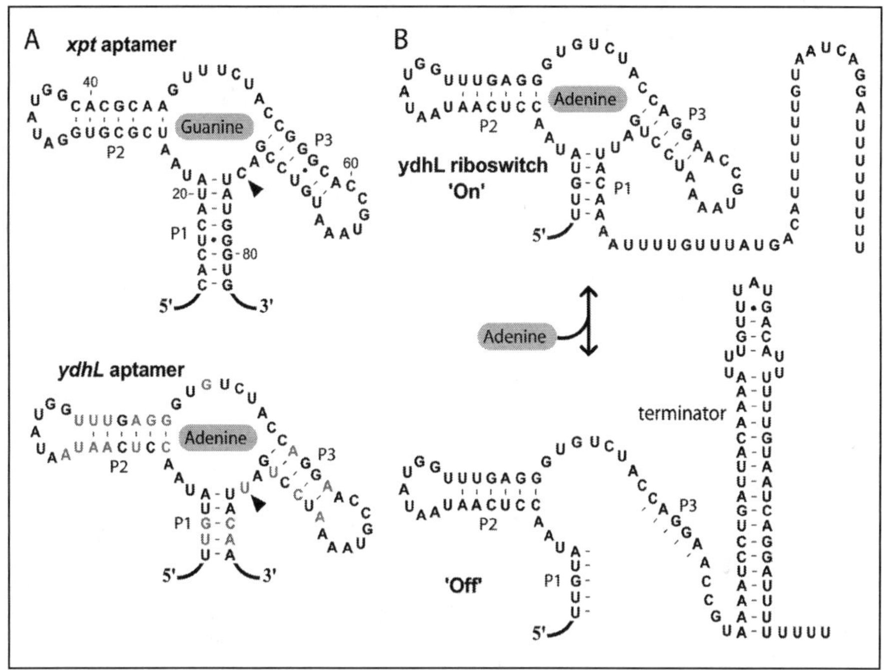

Figure 4. Guanine- and adenine-specific riboswitches. A) The sequences for a natural guanine-binding aptamer (top) and a variant aptamer that binds adenine (bottom), both from *B. subtilis*. Of 23 differences between the two RNAs (shaded nucleotides of *ydhL*), it is most likely that the single C-to-U mutation (arrowhead) determines whether the variants bind guanine or adenine. B) Proposed mechanism for activation of gene expression by an adenine-specific riboswitch. Details are described elsewhere.[60]

A Ribozyme Riboswitch

One of the more sophisticated riboswitches identified to date is also a self-cleaving RNA. The *glmS* ribozyme has a complex secondary structure (Fig. 5A) and carries a substantial number of nucleotides that remain highly conserved (Fig. 2H) throughout Gram-positive organisms.[11] This RNA element, first identified by using bioinformatics, is present only once in bacterial genomes where it is found. It resides immediately upstream of the *glmS* gene encoding the protein enzyme glutamine-fructose-6-phosphate amidotransferase, which uses glutamine and fructose-6-phosphate to generate glucosamine-6-phosphate (GlcN6P).

Upon identification of this motif, it was examined for possible allosteric function with several candidate ligands, including the metabolite product of the GlmS protein encoded downstream. Surprisingly, the RNA from *B. subtilis* exhibits an extraordinarily high level of site-specific cleavage activity that is dramatically accelerated by GlcN6P (Fig. 5B). This 1000-fold acceleration of ribozyme activity causes the RNA to self-cleave with a half-life of ~15 seconds. Furthermore, this natural ribozyme undergoes "rapid switching", wherein the ribozyme adopts an inactive state in the absence of its metabolite trigger, but rapidly converts into an active state upon addition of the ligand. Activation of ribozyme function does not require a forceful denaturation and reannealing treatment, suggesting that the RNA makes use of rapid kinetics to sample alternative states most likely on a seconds or sub-seconds timescale.

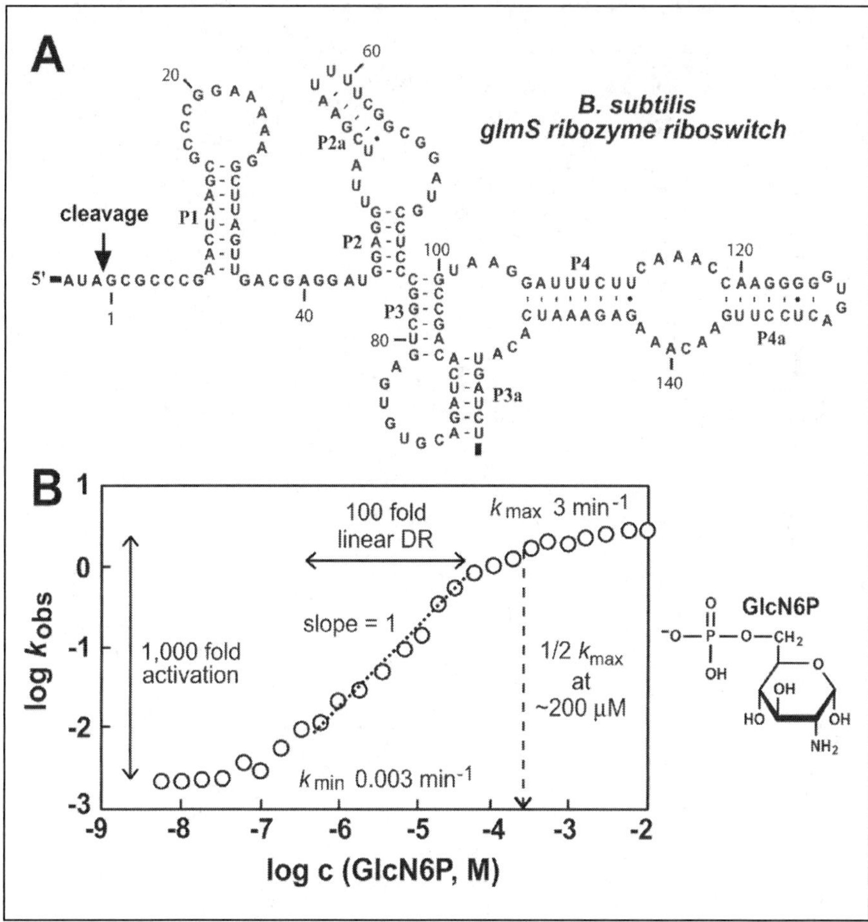

Figure 5. A ribozyme riboswitch. A) Sequence and secondary structure model for the *glmS* ribozyme from *B. subtilis*.[11] Base-paired elements are designated P1 through P4, and the cleavage site is designated by the arrow. B) Kinetic characteristics of metabolite activation by the *glmS* ribozyme. The metabolite effector, glucosamine-6-phosphate (GlcN6P), induces a linear increase in ribozyme activity that approaches its maximum when present at a concentrations exceeding 200 μm. DR indicates the dynamic range for ribozyme activation.

The performance characteristics of this natural ribozyme riboswitch are very similar to the characteristics of allosteric RNAs that were created by molecular engineering. Numerous examples of allosteric hammerhead ribozymes were created first by grafting aptamers or random-sequence domains onto parts of various ribozymes whose structures were critical for activity.[33,34,36-38,64] Although some constructs made by this modular rational design approach already are allosteric, populations of these conjoined RNAs can be subjected to in vitro selection to identify functional variants. The most proficient engineered RNAs of this type exhibit allosteric rate enhancements of ~100,000 fold,[36,65] although many others exhibit a level of allosteric control that is more comparable with the natural *glmS* ribozyme. These results suggest that engineered ribozymes with different ligand sensitivities could be created and made to serve as designer gene control or biosensor elements.

Although RNA cleavage occurs within the 5' UTR and not within the adjoining ORF, this metabolite-induced activity correlates with down-regulation of gene expression. It is not obvious how mRNA cleavage within the 5′ UTR results in reduction of gene expression, but perhaps this event might trigger further degradation. What is also puzzling is the fact that a self-cleaving ribozyme mechanism is used at all. Other riboswitch classes identified to date simply make use of alternatively folded structures that modulate gene expression without inducing a chemical transformation. Why does this particular riboswitch make use of a self-cleaving event, when presumably a simpler form of this motif without catalytic function would likely be sufficient to control expression of the downstream ORF? It is notable that short stretches of nucleotides immediately upstream of the ribozyme cleavage site also are highly conserved but are not required for ribozyme activity. Therefore, this riboswitch might need to be a ribozyme to liberate this additional conserved RNA fragment for possible roles elsewhere in the cell.

A Cooperative Riboswitch

As described above, most riboswitches make use of a simple one-to-one relationship between ligand and RNA, which yields a linear response in gene expression to changing concentrations of ligand. In contrast, many protein genetic factors use multiple polypeptides to generate a more "digital" response to changing ligand concentrations.[66] If two ligand-binding domains interact with perfect cooperativity, then a 10-fold increase in target concentration would yield a 100-fold change in the level of gene expression. Arrangements of this type are essential if dramatic changes in the level of gene expression are required over very limited changes in the concentration of a particular metabolite.

Riboswitches that make use of cooperative ligand binding also exist. A bioinformatics approach has revealed the existence of two similar RNA motifs (type I and type II) that reside upstream of genes encoding for the glycine cleavage system in *B. subtilis* and other *Bacillus* and *Clostridium* bacteria.[67] This protein complex catalyzes the chemical transformations needed to utilize excess glycine as a carbon source. Surprisingly, the 5′ UTR of the *gcvT-gcvPA-gcvPB* operon from *B. subtilis* carries both type I and type II aptamers in close proximity (Fig. 6A). This tandem aptamer configuration is present in almost all other organisms as well, suggesting that this arrangement is important for the function of the riboswitch.

Both biochemical and genetic data indicate that glycine riboswitches carrying tandem aptamers selectively respond to glycine, and that ligand binding induces the riboswitches to activate gene expression. In *B. subtilis*, activation of gene expression appears to result from the fact that the second aptamer forms a ligand-dependent structure that precludes the formation of an intrinsic transcription terminator (Fig. 6A). Furthermore, RNA probing data, equilibrium dialysis experiments, and in vitro transcription assays indicate that the tandem aptamer configuration facilitates the binding of two glycine molecules in a cooperative fashion.[12]

This cooperative binding permits the riboswitch to function as a genetic "ON" switch that is more responsive to changing glycine concentrations (Fig. 6B). Presumably, this allows the cell to respond more substantially than typical riboswitches to small increases in glycine and produce more copies of the glycine cleavage system. Likewise, the tandem aptamer arrangement permits the riboswitch to respond more substantially to small decreases in glycine, so as to prevent this important compound from been depleted to a point where protein synthesis might be compromised.

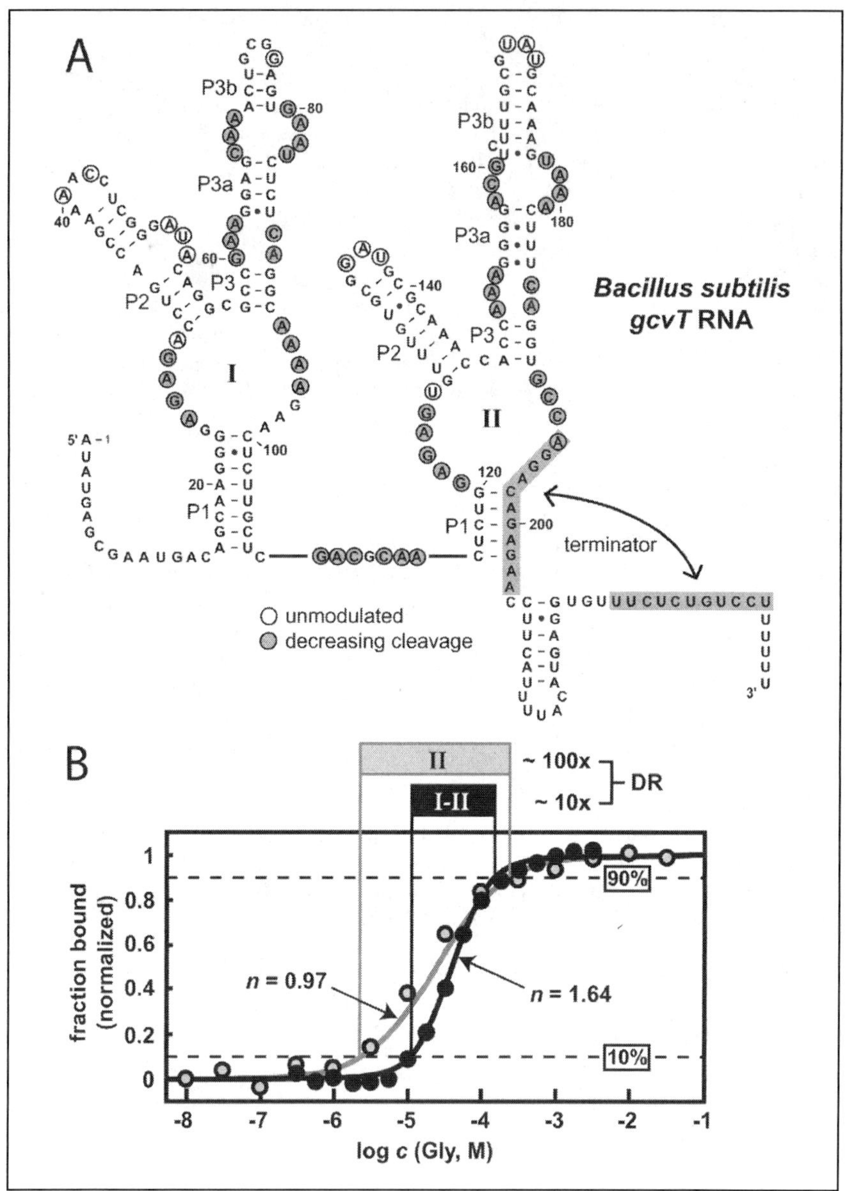

Figure 6. A cooperative glycine-binding riboswitch with tandem aptamers. A) Sequence, secondary structure model, and proposed mechanism for the glycine-specific riboswitch from *B. subtilis*. Upon the introduction of glycine, the two aptamer domains (I and II) exhibit similar changes in their patterns of spontaneous cleavage at internucleotide linkages that reside immediately 3′ of the nucleotides identified by shaded circles. The shaded boxes identify complementary sequences that can form an intrinsic transcription terminator. B) Plot depicting the fraction of RNA bound to ligand versus increasing concentrations of glycine. The dynamic range (DR) of glycine concentrations that result in a change in the form of glycine-bound RNA from 10% to 90% is ~100 fold for an RNA construct carrying a single aptamer (II) and is ~10 fold for an RNA construct that carries both aptamers (I-II). Additional details are described elsewhere.[12]

Interestingly, a similar tandem aptamer arrangement was created for other ligands using molecular engineering. This engineered construct is composed of five functional elements: a hammerhead ribozyme, aptamers for FMN and theophylline, and two bridging domains that both link the multi-domain construct together and permit allosteric activity. The binding of theophylline to its corresponding aptamer improves the affinity of the FMN aptamer for its ligand by more than 80-fold.[68] Similarly, the binding of glycine to either aptamer within the natural riboswitch improves the binding affinity in the second site by ~100-fold or greater. Furthermore, the degree of cooperativity exhibited by the natural riboswitch compares favorably with that of allosteric proteins. These findings from both RNA engineering studies and from studies on natural riboswitches demonstrate that RNA can form genetic switches with a level of structural and functional sophistication that approaches that of protein genetic factors.

Finding New Riboswitches

Although the number of riboswitch classes identified already exceeds the number of natural ribozyme classes that have been discovered, the full scope of genetic control by riboswitches has yet to be established. This goal could be attained quickly if methods for the rapid identification of novel RNA motifs were established. Clues to the existence of several riboswitches have been provided by genetic experiments conducted over the last 30 years, and these early studies mostly centered on what we now know to be the most common classes of riboswitches. Specifically, riboswitches for TPP, coenzyme B_{12}, FMN, SAM, and lysine are common among some of the best-studied bacterial lineages, and thus the probability that genetics studies would have encountered them was high. In contrast, some riboswitches occur less frequently, and data regarding their structures and functions are not present in the literature.

The availability of numerous genomic DNA sequences for bacteria provides an alternative means by which new riboswitches can be identified. An efficient strategy to identify new highly conserved elements involves searching for sequence homology exclusively within noncoding portions of genomic DNA. A combination of sequence homology searching, phylogenetic sequence alignment, and secondary structure prediction has proven to be a productive route to the discovery of novel riboswitch classes and novel RNAs that might function as riboswitches.

The initial bioinformatics approaches used to identify new riboswitch candidates rely on the fact that at least some riboswitches are large and remain exceptionally well conserved through evolution (Fig. 7). Moreover, the DNA templates for riboswitches typically reside in intergenic regions (IGRs) immediately upstream and are templated by the same strand of DNA as are the protein-coding ORFs that they control. Therefore, new riboswitches are likely to be identified by using computer algorithms to identify stretches of sequence homology residing within large IGRs from numerous bacterial genomes. Further examination of hits for a given sequence class yields a consensus sequence and a predicted secondary structure model for newly identified riboswitch candidates.

In one study using DNA sequence data from 91 complete bacterial genomes, we identified eight new sequence elements that have characteristics of riboswitches.[67] Specifically, each element resides upstream of genes that in most cases appear to be related in biochemical function. Each element exhibits conservation of sequence domains and base-paired structures. Moreover, transcribed RNAs from each element exhibit signs of forming complex structures.

It is possible that some of these new motifs might not function as metabolite-binding riboswitches, but could represent protein recognition sites or could function in some other capacity as a noncoding RNA. However, such RNA motifs are usually not as widely distributed

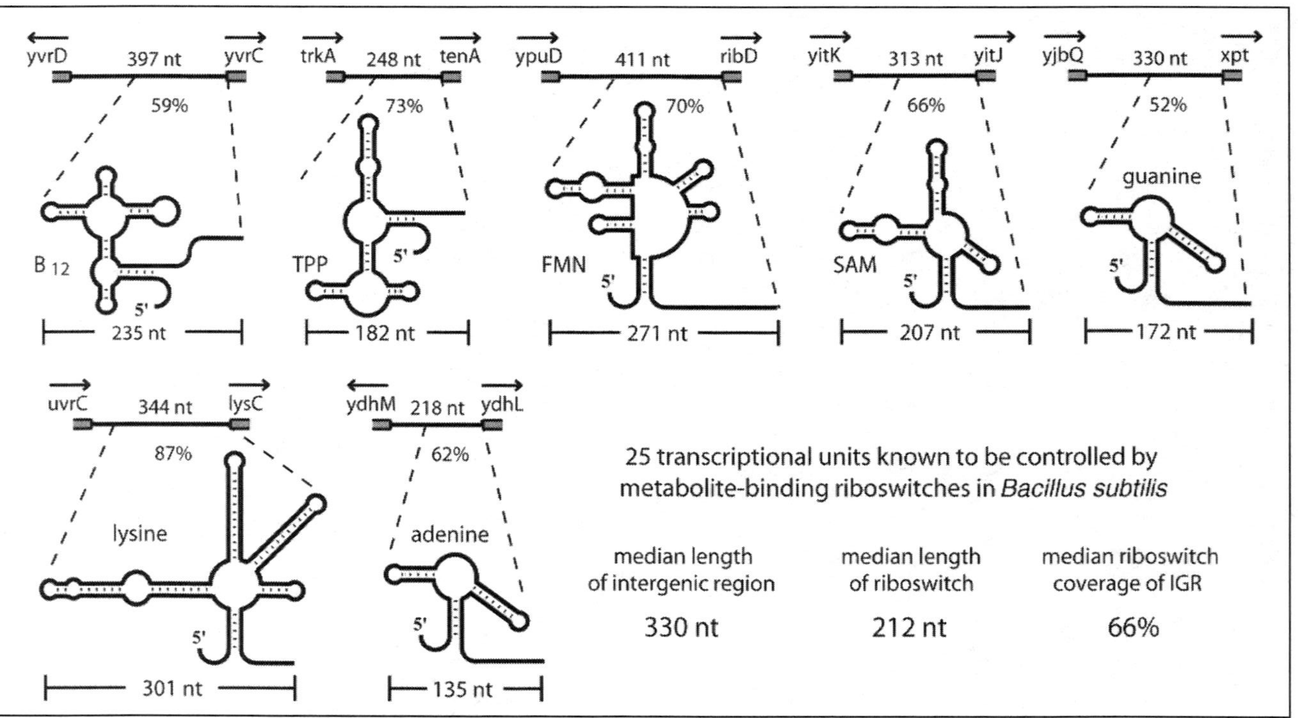

Figure 7. Genomic characteristics of bacterial riboswitches. Depicted are single representatives of seven classes of riboswitches that are common in *Bacillus* and *Clostridium* bacteria. For each example, the genomic location is indicated by defining the flanking genes (shaded boxes). The number of nucleotides (nt) for each *B. subtilis* IGR (thin line) is provided, as is the number of nucleotides present in each riboswitch (from the start of the aptamer domain to the AUG start codon for the ORF). Percentages reflect the amount of each IGR that is used as a template for producing the riboswitch. The median values for IGR length, riboswitch length, and percent coverage by riboswitches was calculated for 25 riboswitches of these seven classes from *B. subtilis*.

among bacterial lineages, or are not as well conserved in sequence and secondary structure as are the riboswitches from the known classes. It is important to note that not all riboswitches are sufficiently large, well-conserved, or widespread to be readily identified by existing bioinformatics algorithms. Small or highly variable riboswitches, or those that might be only poorly distributed among bacteria, will be more difficult to recognize by computational means.

Are Riboswitches Ancient?

A larger role for RNA is precisely what is required to provide a realistic foundation for more robust versions of the RNA World hypothesis.[69] If life passed through an age wherein all functions were carried out by RNA, and if this age gave rise to complex organisms with a diversity of ribozymes and metabolic pathways, then there must have been mechanisms present that permitted these ribo-organisms to detect and respond to various chemical cues.[70] Of course, establishing the precise composition and functions of ancient life forms is extraordinarily difficult. However, researchers can begin to establish the functional limitations of nucleic acids by using RNA engineering strategies and by examining the roles of RNA in modern cellular life. In this way, the potential for sophisticated function by RNA can be coupled to evolutionary history, such that a plausible framework can be constructed to help visualize the biochemistry of ancient organisms.

Although the organisms that represented the best of purely RNA World life have all become extinct, some of their biochemical capabilities might still be exploited by modern cells. The ribosome is the most striking example of an RNA enzyme whose origin almost assuredly predates the rise of proteins. This RNA-protein complex that synthesizes all encoded proteins carries an RNA enzyme at its peptidyl-transferase core.[71-73] Perhaps even more pieces of ancient molecular machinery from the RNA World remain with us today, where they carry out functions that are nearly the same as they were several billion years ago.

A more loosely controlled metabolic state could have been guided by the activities of ribozymes themselves. Parameters such as the rate constants for various ribozymes, their copy number, and the concentrations of ribozyme substrates would have been important in establishing the metabolic flux through primitive RNA World biosynthetic pathways. Certainly, the presence of riboswitch-like structures would have enhanced the efficiency of metabolic processes in organisms of the RNA World. Obviously, these first riboswitches would not have been controlling the expression of protein enzymes, but they would have controlled the production, processing and activity of their ribozyme counterparts. The *glmS* ribozyme is one modern example of metabolite-mediated ribozyme control of RNA processing.

Intriguingly, it is possible that some of the riboswitches discovered to date are close relatives of RNA World metabolite sensors. So far, all of the metabolites known to be sensed by riboswitches are of fundamental importance to nearly all extant organisms. Admittedly, this could be due to the fact that cells need to sense the concentrations of such important compounds and that RNA has been chosen for this task more recently in evolution, despite competition from protein factors. Certainly, the characteristics of at least some riboswitches (e.g., wide phylogenetic distribution and binding of putative RNA World coenzymes) are precisely as expected if they had emerged from an RNA-centric metabolic state.[69,74-76]

Perspective

The existence of riboswitches is intriguing for a variety of reasons. Riboswitches are natural proof for the concept that RNA has the structural and functional diversity needed to bind chemical targets with high affinity and specificity. The fact that riboswitches are entrusted to sense compounds that are fundamental to nearly all organisms suggests that molecular switches made of RNA have the functional sophistication needed to be competitive with proteins in evolution. Furthermore, riboswitches are not just static receptors for their targets, but also take an active role in affecting the cellular machinery involved in expressing genes. Because a single messenger RNA encompasses individual elements that act as a molecular sensor, a genetic switch, and a coding region, riboswitches provide the cell with a mechanism for gene control that in some instances is far simpler than that presented by mechanisms requiring protein factors.

It is interesting to note that this simpler means of genetic control permits RNA to perform a major function likely to be required by any life form that needs to maintain a complex metabolic state. Thus, riboswitches are precisely what is required of RNA if organisms of the RNA World were carrying out sophisticated metabolic pathways. Perhaps not surprising then are the observations that some characteristics of riboswitches are consistent with their early emergence in evolution, and therefore they could represent an ancient form of metabolite sensing and biochemical control system.

Finally, it is relatively straightforward to manipulate RNA structural elements. RNA structures, including the aptamer domains from riboswitches, can be highly modular. Therefore, it is possible that variant RNA constructs could be created to serve as novel riboswitches.[77] In addition, combinatorial strategies such as in vitro evolution could be used to further manipulate and refine the functions of engineered riboswitches. Thus RNA engineering strategies could give rise to riboswitches that carry designer ligand-sensing elements, which would be useful for creating novel gene control networks.

References

1. Nahvi A, Sudarsan N, Ebert MS et al. Genetic control by a metabolite binding mRNA. Chem Biol 2002; 9:1043-1049.
2. Winkler W, Nahvi A, Breaker RR. Thiamine derivatives bind messenger RNAs directly to regulate bacterial gene expression. Nature 2002; 419:952-956.
3. Lai EC. RNA sensors and riboswitches: self-regulating messages. Curr Biol 2003; 13:285-291.
4. Sudarsan N, Barrick JE, Breaker RR. Metabolite-binding RNA domains are present in the genes of eukaryotes. RNA 2003; 9:644-647.
5. Muller S. Another face of RNA: metabolite-induced "riboswitching" for regulation of gene expression. ChemBioChem 2003; 4:817-819.
6. Winkler WC, Breaker RR. Genetic control by metabolite-binding riboswitches. ChemBioChem 2003; 4:1024-1032.
7. Nudler E, Mironov AS. The riboswitch control of bacterial metabolism. Trends Biochem Sci 2004; 29:11-17.
8. Vitreschak AG, Rodionov DA, Mironov AA et al. Riboswitches: the oldest mechanism for the regulation of gene expression? Trends Genet 2004; 20:44-50.
9. Soukup JK, Soukup GA. Riboswitches exert genetic control through metabolite-induced conformational change. Curr Opin Struct Biol 2004; 14:344-349.
10. Mandal M, Boese B, Barrick JE et al. Riboswitches control fundamental biochemical pathways in Bacillus subtilis and other bacteria. Cell 2003; 113:577-586.
11. Winkler WC, Nahvi A, Roth A et al. Control of gene expression by a natural metabolite-responsive ribozyme. Nature 2004; 428:281-286.
12. Mandal M, Lee M, Barrick JE et al. A glycine-dependent riboswitch that uses cooperative binding to control gene expression. Science 2004; 306:275-279.
13. Winkler WC, Roth A, Collins JA et al. (manuscript in preparation).
14. Corbino KA, Narasimhan S, Weinberg Z et al. (manuscript in preparation).
15. Griffiths-Jones S, Bateman A, Marshall M et al. Rfam: an RNA family database. Nucleic Acids Res 2003; 31:439-441.
16. Tang J, Breaker RR. Rational design of allosteric ribozymes. Chem Biol 1997; 4:453-459.
17. Jacob F, Monod J. Genetic regulatory mechanisms in the synthesis of proteins. J Mol Biol 1961; 3:318-356.
18. Cech TR. Ribozymes, the first 20 years. Biochem Soc Trans 2002; 30:1162-1166.
19. Doudna JA, Cech TR. The chemical repertoire of natural ribozymes. Nature 2002; 418:222-228.

20. Gold L, Brown D, He Y et al. From oligonucleotide shapes to genomic SELEX: novel biological regulatory loops. Proc Natl Acad Sci USA 1997; 94:59-64.
21. Stormo GD, Ji Y. Do mRNAs act as direct sensors of small molecules to control their expression? Proc Natl Acad Sci USA 2001; 98:9465-9467.
22. Yanofsky C. Attenuation in the control of expression of bacterial operons. Nature 1981; 289:751-758.
23. Gusarov I, Nudler E. The mechanism of intrinsic transcription termination. Mol Cell 1999; 3:495-504.
24. Yarnell WS, Roberts JW. Mechanism of intrinsic transcription termination and antitermination. Science 1999; 284: 611-615.
25. Grundy FJ, Henkin TM. The T box and S box transcription termination control systems. Front Biosci 2003; 8:20-31.
26. Morita MT, Tanaka Y, Kodama TS et al. Translational induction of heat shock transcription factor sigma32: evidence for a built-in RNA thermosensor. Genes Dev 1999; 13:655-665.
27. Johansson J, Mandin P, Renzoni A et al. An RNA thermosensor controls expression of virulence genes in Listeria monocytogenes. Cell 2002; 110:551-561.
28. Chowdhury S, Ragaz C, Kreuger E et al. Temperature-controlled structural alterations of an RNA thermometer. J Biol Chem 2003; 278:47915-47921.
29. Gold L, Polisky B, Uhlenbeck O et al. Diversity of oligonucleotide functions. Annu Rev Biochem 1995; 64:763-797.
30. Osborne SE, Ellington AD. Nucleic acid selection and the challenge of combinatorial chemistry. Chem Rev 1997; 97:349-370.
31. Hermann T, Patel DJ. Adaptive recognition by nucleic acid aptamers. Science 2000; 287:820-825.
32. Jenison RD, Gill SC, Pardi A et al. High-resolution molecular discrimination by RNA. Science 1994; 263:1425-1429.
33. Soukup GA, Emilsson GA, Breaker RR. Altering molecular recognition of RNA aptamers by allosteric selection. J Mol Biol 2000; 298:623-632.
34. Soukup GA, Breaker RR. Engineering precision RNA molecular switches. Proc Natl Acad Sci USA 1999; 96:3584-3589.
35. Robertson MP, Ellington AD. Design and optimization of effector activated ribozyme ligases. Nucleic Acids Res 2000; 28:1751-1759.
36. Seetharaman S, Zivarts M, Sudarsan N et al. Immobilized RNA switches for the analysis of complex chemical and biological mixtures. Nat Biotechnol 2001; 19:336-341.
37. Breaker RR. Engineered allosteric ribozymes as biosensor components. Curr Opin Biotechnol 2002; 13:31-39.
38. Silverman SK. Rube Goldberg goes (ribo)nuclear? Molecular switches and sensors made from RNA. RNA 2003; 9:377-383.
39. Lundrigan MD, Koster W, Kadner RJ. Transcribed sequences of the Escherichia coli btuB gene control its expression and regulation by vitamin B_{12}. Proc Natl Acad Sci USA 1991; 88:1479-1483.
40. Ravnum S, Andersson DI. Vitamin B_{12} repression of the btuB gene in Salmonella typhimurium is mediated via a translational control which requires leader and coding sequences. Mol Microbiol 1997; 23:35-42.
41. Richter-Dahlfors AA, Ravnum S, Andersson DI. Vitamin B_{12} repression of the cob operon in Salmonella typhimurium: translational control of the cbiA gene. Mol Microbiol 1994; 13:541-553.
42. Nou X, Kadner RJ. Adenosylcobalamin inhibits ribosome binding to btuB RNA. Proc Natl Acad Sci USA 2000; 97:7190-7195.
43. Ravnum S, Andersson DI. An adenosyl-cobalamin (coenzyme-B_{12})-repressed translational enhancer in the cob mRNA of Salmonella typhimurium. Mol Microbiol 2001; 39:1585-1594.
44. Gelfand MS, Mironov AA, Jomantas J et al. A conserved RNA structure element involved in the regulation of bacterial riboflavin synthesis genes. Trends Genet 1999; 15:439-442.
45. Mironov VN, Perumov DA, Kraev AS et al. Unusual structure of the regulatory region of the riboflavin biosynthesis operon in Bacillus subtilis. Mol Biol (Mosk) 1990; 24:256-261.
46. Kreneva RA, Perumov DA. Genetic mapping of regulatory mutations of Bacillus subtilis riboflavin operon. Mol Gen Genet 1990; 222:467-469.
47. Kil YV, Mironov VN, Gorishin I et al. Riboflavin operon of Bacillus subtilis: unusual symmetric arrangement of the regulatory region. Mol Gen Genet 1992; 233:483-486.

48. Webb E, Febres F, Downs DM. Thiamine pyrophosphate (TPP) negatively regulates transcription of some thi genes of Salmonella typhimurium. J Bacteriol 1996; 178:2533-2538.
49. Miranda-Rios J, Navarro M, Soberon M. A conserved RNA structure (thi box) is involved in regulation of thiamin biosynthetic gene expression in bacteria. Proc Natl Acad Sci USA 2001; 98:9736-9741.
50. Mironov AS, Gusarov I, Rafikov R et al. Sensing small molecules by nascent RNA: a mechanism to control transcription in bacteria. Cell 2002; 111:747-756.
51. Winkler WC, Cohen-Chalamish S, Breaker RR. An mRNA structure that controls gene expression by binding FMN. Proc Natl Acad Sci USA 2002; 99:15908-15913.
52. Mandal M, Breaker RR. Gene regulation by riboswitches. Nat Rev Mol Cell Biol 2004; 5:451-463.
53. McDaniel BA, Grundy FJ, Artsimovitch I et al. Transcription termination control of the S box system: direct measurement of S-adenosylmethionine by the leader RNA. Proc Natl Acad Sci USA 2003; 100:3083-3088.
54. Epshtein V, Mironov AS, Nudler E. The riboswitch-mediated control of sulfur metabolism in bacteria. Proc Natl Acad Sci USA 2003; 100:5052-5056.
55. Winkler WC, Nahvi A, Sudarsan N et al. An mRNA structure that controls gene expression by binding S-adenosylmethionine. Nat Struct Biol 2003; 10:701-707.
56. Grundy FJ, Lehman SC, Henkin TM. The L box regulon: lysine sensing by leader RNAs of bacterial lysine biosynthesis genes. Proc Natl Acad Sci USA 2003; 100:12057-12062.
57. Sudarsan N, Wickiser JK, Nakamura S et al. An mRNA structure in bacteria that controls gene expression by binding lysine. Genes Dev 2003; 17:2688-2697.
58. Nahvi A, Barrick JE, Breaker RR. Coenzyme B_{12} riboswitches are widespread genetic control elements in prokaryotes. Nucleic Acids Res 2004; 32:143-150.
59. Vitreschak AG, Rodionov DA, Mironov AA et al. Regulation of the vitamin B_{12} metabolism and transport in bacteria by a conserved RNA structural element. RNA 2003; 9:1084-1097.
60. Mandal M, Breaker RR. Adenine riboswitches and gene activation by disruption of a transcription terminator. Nat Struct Mol Biol 2004; 11:29-35.
61. Batey RT, Gilbert SD, Montange RK. Structure of a natural guanine-responsive riboswitch complexed with the metabolite hypoxanthine. Nature 2004; 432:411-415.
62. Serganov A, Yuan Y, Pikovskaya O et al. Structural basis for discriminative regulation of gene expression by adenine- and guanine-sensing mRNAs. Chem Biol 2004; 11:1729-1741.
63. Wickiser JK, Winkler WC, Breaker RR et al. The speed of RNA transcription and metabolite binding kinetics operate an FMN riboswitch. Mol Cell 2005; 18:49-60.
64. Koizumi M, Soukup GA, Kerr JN et al. Allosteric selection of ribozymes that respond to the second messengers cGMP and cAMP. Nat Struct Biol 1999; 6:1062-1071.
65. Robertson MP, Ellington AD. In vitro selection of nucleoprotein enzymes. Nat Biotechnol 2001; 19:650-655.
66. Ptashne M, Gann A. Genes & Signals. Cold Spring Harbor: Cold Spring Harbor Laboratory Press, 2002.
67. Barrick JE, Corbino KA, Winkler WC et al. New RNA motifs suggest an expanded scope for riboswitches in bacterial genetic control. Proc Natl Acad Sci USA 2004; 101:6421-6426.
68. Jose AM, Soukup GA, Breaker RR. Cooperative binding of effectors by an allosteric ribozyme. Nucleic Acids Res 2001; 29:1631-1637.
69. Benner SA, Ellington AD, Tauer A. Modern metabolism as a palimpsest of the RNA world. Proc Natl Acad Sci USA 1989; 86:7054-7058.
70. Joyce GF. The antiquity of RNA-based evolution. Nature 2002; 418:214-221.
71. Ban N, Nissen P, Hansen J et al. The complete atomic structure of the large ribosomal subunit at 2.4 Å resolution. Science 2000; 289:905-920.
72. Nissen P, Hansen J, Ban N et al. The structural basis of ribosome activity in peptide bond synthesis. Science 2000; 289:920-930.
73. Cech TR. Structural biology. The ribosome is a ribozyme. Science 2000; 289:878-879.
74. White 3rd HB. Coenzymes as fossils of an earlier metabolic state. J Mol Evol 1976; 7:101-104.
75. Jeffares DC, Poole AM, Penny D. Relics from the RNA world. J Mol Evol 1998; 46:18-36.
76. Jadhav VR, Yarus M. Coenzymes as coribozymes. Biochimie 2002; 84:877-888.
77. Breaker RR. Natural and engineered nucleic acids as tools to explore biology. Nature 2004; 432:838-845.

CHAPTER 7

Switchable RNA Motifs as Drug Targets

Eric Westhof,* Boris François and Quentin Vicens

Abstract

RNA molecules are highly negatively charged polymers that form intricate three-dimensional assemblies involving recurrent structural motifs. Therefore, in order to understand the molecular recognition of RNA, one of the key points to address is how RNA can be a specific target of natural or artificial antibiotics and drugs that are generally positively charged. Crystal structures of complexes between ribosomal particles from bacteria and antibiotics have pinned down very precisely the discrete binding sites of several classes of antibiotics that inhibit protein synthesis. These structures have unambiguously demonstrated that ribosomal RNAs, rather than ribosomal proteins, are overwhelmingly targeted. The comparative analyses of various aminoglycoside antibiotics bound to the same aminoacyl-transfer RNA (tRNA) decoding site (A site) have been used to decipher the contribution of each functional group to the RNA-aminoglycoside complex formation. In addition, various biochemical and microbiological data as well as some resistance and toxicity mechanisms could be rationalized at the molecular level. It was demonstrated that the binding of the aminoglycosides locks the A site into a conformation mimicking that adopted in presence of the cognate tRNA-codon association, thereby provoking a loss in translation fidelity by shunting a natural molecular switch. Similarly, although very high specificity might be difficult to achieve with oppositely charged molecules, targeting motifs that undergo dynamic exchange between alternative conformations (molecular switches) should improve the biological activity of antibacterial compounds.

Introduction

The ribosome is the target of about half of the antibiotics characterized thus far.[1-3] For forty years, microbiological, pharmacological, and biochemical data have helped to decipher the mechanisms of action of various antibiotics, by providing clues about their binding sites (e.g., through footprinting experiments[4] and via identification of mutations[5]) as well as their mechanisms of action (e.g., by kinetic measurements[6]). A critical advance in the understanding of these mechanisms has recently been made with the high-resolution crystal structures (2.4 - 3.8 Å) of bacterial ribosomal particles complexed to several classes of protein synthesis inhibitors (aminoglycosides, macrolides, chloramphenicol, etc.).[7-12] These structures definitively show that antibiotics predominantly target ribosomal RNA molecules rather than ribosomal proteins. Crystal structures have also been solved (2.4 - 2.54 Å) for the A site, an

*Corresponding Author: Eric Westhof—Institut de Biologie Moléculaire et Cellulaire du CNRS, Modélisation et Simulations des Acides Nucléiques, UPR 9002, Université Louis Pasteur, 15 rue René Descartes, 67084 Strasbourg Cedex, France. Email: E.Westhof@ibmc.u-strasbg.fr

Nucleic Acid Switches and Sensors, edited by Scott K. Silverman. ©2006 Landes Bioscience and Springer Science+Business Media.

isolated domain of the 16S ribosomal RNA, in complex with several antibiotics from the aminoglycoside family (paromomycin, tobramycin and geneticin).[13-15] The comparative analyses of these high-resolution structures aided in deciphering the contribution of each antibiotic functional group to the binding, and they offered a molecular basis to explain some resistance and toxicity mechanisms.[16]

Here, starting from the challenges that small molecules must face to target RNA and with our present understanding of the molecular recognition between rRNA and aminoglycosides used as a model system, we will stress the advantages of targeting RNA molecular switches, whose recurrence is now revealed in various RNA molecules. Some of those aspects were discussed in a previous minireview.[17]

Constraints of RNA Folding on the Choice of RNA Target

Experimental and theoretical studies on the three-dimensional architecture of catalytic RNAs revealed the hierarchical folding of structured RNAs.[18-20] Secondary structure pairings join regions that are proximate in sequence, and these secondary structure elements subsequently stack end-to-end to form contiguous helices. Such preformed helical domains associate into bundles of helices to constitute the compact tertiary structure that is maintained via interactions between tertiary anchoring motifs.[21] Thus, RNA architecture can be visualized as the hierarchical assembly of preformed double-stranded helices defined by Watson-Crick base pairs and RNA modules principally maintained by nonWatson-Crick base pairs. This architectural hierarchy is coupled with an electrostatic hierarchy in which RNA folding occurs first with an electrostatic collapse to compact states, with most of the secondary structure elements induced by nonspecific ion binding.[22] Later, there is a cooperative transition to native states, with all tertiary contacts induced by specific ion binding, especially magnesium ions.[23] Similar arrangements of domains occur in smaller systems like the hammerhead,[24,25] the hairpin,[26] and the hepatitis delta virus ribozymes,[27,28] and some of these RNA motifs are also found in the large ribosomal RNAs.[29]

Therefore, although RNA does not appear to be a very promising drug target from its chemical structure (built on only four different kinds of negatively charged nucleotides possessing planar bases[30]), one could argue that the intricate architectures of RNA molecules can still lead to the formation of pockets and cavities where shape-specific rather than sequence-specific binding could be achieved.[31] Several observations may be made regarding this argument.

First, the formation of RNA cavities necessitates turns of the sugar-phosphate backbone and thus a close proximity of phosphate groups. This leads to a heightened importance of electrostatic forces and increases the roles of tightly bound water molecules and ions that screen the repulsive charges. Divalent magnesium ions need special considerations. Although the roles of monovalent ions cannot be dismissed, magnesium ions are generally necessary for RNA to fold and function. In order to bind specifically to a geometrically restrained and molecularly crowded pocket, they must be partly dehydrated, which is energetically very costly because the enthalpy of hydration of a single magnesium ion is around 400 kcal/mole. The displacement of such tightly bound ions by another positively charged ligand, with the accompanying structural rearrangements, is thus not favorable.

Second, the formation of pockets or enlarged grooves requires the presence of non-Watson-Crick pairs and bulged residues. The associated sequence constraints are usually rather strong and the number of ways of embedding non-Watson-Crick pairs within helices is limited, which leads to a rather restricted number of RNA motifs that could be chosen as potential targets.[29,32] Indeed, RNA motifs strikingly appear like Russian dolls with smaller motifs associated into larger motifs.[33] Biological function arises through the very diverse architectures that can result from the assembly of those recurrent but limited RNA motifs. The modes of interaction between these motifs are very similar if not identical in the various architectures. The targeting of such RNA-RNA anchor motifs cannot readily lead to a drug that is specific for a particular site. Besides, this would require competition for binding during the folding and assembly processes, which normally involve several protein cofactors with possible ATP or GTP hydrolysis.

Finally, most of the energy content of a given folded RNA is contained in the secondary structure, which consists of regular double-stranded Watson-Crick paired helices. As in proteins, the free energy content of a three-dimensional RNA fold is between -5 and -10 kcal/mole.[23] Thus, a binding constant in the nanomolar range—which can be achieved by a small molecule—could in principle compete with the final steps of RNA folding. However, it should be kept in mind that the folding free energy is distributed throughout the RNA molecule and is not localized in a single region or interface, as is the binding free energy of a small ligand.

Which RNA Dynamics Should Be Targeted?

Like all molecules, RNA molecules spontaneously undergo dynamic motions and movements, the amplitude and frequency of which depend on the temperature (kT). Such Brownian motions are expected to lead to localized alternative conformational states of a folded RNA molecule. High resolution X-ray crystallography reveals such alternative conformations of the sugar-phosphate backbone. A particular example, relevant to the present topic, is seen in the crystal structure of the complex between the aminoglycoside geneticin and the A-site RNA fragment.[15] The types of mobilities extend from base rotations about the sugar or localized conformational fluctuations to disorder. Thus, mobility generally covers atomic, structural, segmental, or domain movements. Clearly, biological functional significance cannot be systematically assigned to these various types of dynamic processes. In proteins, hinge flexibility in immunoglobulins as well as the role of short-range fluctuations in the diffusion of oxygen in myoglobin have been documented, and the flexibility in both the ligands and protein binding sites have been stressed.[34] More often than not, only correlations between segmental mobility and biological function can be found, as was the case for example with antigen-antibody recognition.[35,36] During recognition and binding processes, fluctuations facilitate the interplay of the various physicochemical forces in the search for a minimum in free energy of binding. In other words, the fluctuations let the constraints potentially present in the attractive forces exercise their action effectively for the desolvation step and for the zipper-like propagation of the initial nucleation complex into the lock-and-key tight binding stage. A region targeted for drug binding should thus possess enough internal dynamics and undergo enough atomic fluctuations to allow these accommodation processes during complexation.

RNA molecules experience a vast range of alternative conformations. Dynamic equilibria with correlations between the conformers were noticed several years ago using NMR spectroscopy; e.g., in nucleosides.[37] A single bulged base can occupy various positions within the helical grooves as well as when flipped outside in the solvent, depending on local sequence and conditions.[38,39] Conditions leading to dramatic effects for the resulting tertiary fold can induce rearrangement of the secondary structure. For example, the S15 mRNA exists in an equilibrium between a state with two adjacent hairpins and a state in which these two hairpins have rearranged to a coaxially stacked pseudoknot.[40,41] Biologically, this equilibrium constitutes an important translational control for the S15 protein.

Equilibria between alternative secondary structures are often observed after in vitro transcription of RNA molecules, which results in kinetic trapping of biologically nonnative conformers. This was particularly well analyzed in the central P3/P7 pseudoknot of group I introns.[42-44] Four-way junctions constitute a different and fascinating system for studying topological equilibria. Such equilibria occur because the two-by-two coaxially stacked helices of the four-way junction can form a parallel or antiparallel X-like structure with either a right-handed or a left-handed chirality.[45] Recently, dynamic equilibria beween parallel and antiparallel conformers[46] as well as between right- and left-handed conformers were observed.[47] Similarly, a dynamic equilibrium between an extended and a bent structure has recently been observed[48] for the recurrent kink-turn motif.[49]

What is the difference, if any, between a dynamic equilibrium and a switchable structure? Formally, it is difficult to find one. We consider here that a switchable RNA structure experiences a dynamic equilibrium between at least two main and conformationally characterizable states, such that biological function can be assigned to the existence of the switch. Thus, inherent structural fluctuations (on the order of kT) contributing to RNA folding and ligand recognition are not considered functional RNA switches. This distinction between a dynamic equilibrium and a switchable structure implicitly includes the kinetics of the conformational change. Intuitively, one expects that a dynamic equilibrium has relatively fast exchange kinetics, whereas a switchable structure has relatively slow kinetics. For example, dynamic equilibria that affect ligand binding through allosteric effects do not always lead to conformationally characterizable molecular states, although they are certainly biologically relevant.

Mechanisms of Action of Aminoglycosides

Antibiotics belonging to different families target various regions of ribosomal RNA. They bind in the shallow groove (spectinomycin)[8] or the deep groove (hygromycin B)[7] of a helix, at a three-adenine bulge (aminoglycosides),[8,13] or in the exit tunnel of the nascent polypeptide chain (macrolides).[10,12] Additionally, they interact in many ways with RNA: (i) only with phosphate groups (streptomycin),[8] (ii) mainly with bases (hygromycin B, spectinomycin),[7,8] (iii) with a mixture of both (paromomycin, tobramycin),[8,13,14] (iv) via magnesium ions (tetracycline, chloramphenicol, sparsomycin),[11,12,50] or (v) with a protein side chain (streptomycin).[8] The antibiotics can mimic base stacking (pactamycin)[7] or form pseudo-base pairing interactions with ribosomal bases (blasticidin S, paromomycin, and related aminoglycosides).[8,13,14,50] For clarity, here we concentrate on the aminoglycosides that bind to the decoding A site of the small ribosomal subunit.

Aminoglycoside antibiotics are oligosaccharides that contain several ammonium groups.[51,52] Different sub-classes are distinguished on the basis of their chemical structures (Fig. 1) and their mechanisms of action.[53] Aminoglycosides belonging to the paromomycin and tobramycin sub-classes interfere with translation[54] by binding to the A site on the 16S rRNA (Fig. 2).[4]

Kinetic analyses showed that during decoding, a correct tRNA-mRNA interaction induces a conformational change of the A site that permits translation.[55] Aminoglycosides disturb the fidelity of this tRNA selection step by stabilizing a similar conformation for near-cognate complexes.[6,56-58] The cognate tRNA is the single tRNA species with an anticodon that is precisely complementary to the codon exposed at the A site and is characterized by 1–2 GTP cleavages in the EF-Tu ternary complex per incorporation. Near-cognate aa-tRNAs possess an anticodon similar to the cognate tRNA (4–6 tRNA species) and are characterized by 3–6 GTP cleavages per incorporation. The 90% of noncognate aa-tRNAs that have an anticodon dissimilar to the cognate tRNA are never misincorporated and therefore do not show any GTP cleavages. Kinetic analyses have demonstrated that paromomycin stabilizes binding of both cognate and near-cognate aa-tRNAs in the A site and that GTP hydrolysis and peptide bond formation are accelerated by a factor of ten for the near-cognate complexes.[6] These effects lead to an increase in the level of near-cognate amino acid misincorporation. Simultaneously, paromomycin decreases directly the rate of codon recognition by a factor of three. The latter observation reinforces the model in which aminoglycosides provoke a structural rearrangement of

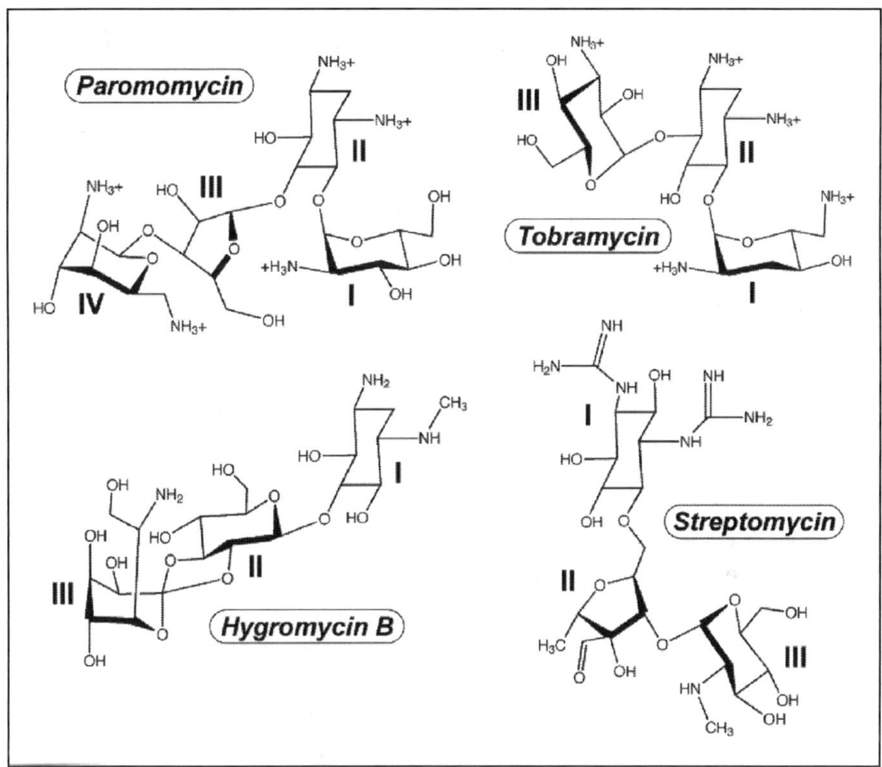

Figure 1. Chemical structures of representatives of common aminoglycosides that belong to four sub-classes.

the decoding site from a state in which it accepts the tRNA to a conformation which is productive for peptide bond synthesis.[58] Recently, by monitoring the rates of GTP hydrolysis by EF-Tu, it could be shown that the conformational steps affected by two aminoglycosides, streptomycin and paromomycin, that bind in close proximity on the 30S particle could be discriminated.[59] In contrast to streptomycin, in the presence of which the rates for GTP hydrolysis are decreased to similar values for cognate and noncognate codons, the rates for GTP hydrolysis are increased for noncognate codons without altering those for cognate codons in presence of paromomycin. Thus, paromomycin binding affects an early tRNA selection step, the specific recognition of the codon-anticodon interaction, whereas streptomycin affects the subsequent triggering of a conformational change.

Aminoglycosides have been shown to inhibit tRNA binding to the A site when the E site is occupied during the elongation cycle.[60] Binding of aminoglycosides would thus lead to a blockage of ribosomal function. Support for the effect of E-site occupation on A-site fidelity was further indicated by the binding of edeine to the E site, which leads to misreading levels comparable to those observed with aminoglycosides.[61] Recently, it was also shown that aminoglycosides inhibit ribosomal subunit formation in *E. coli*[62] and *S. aureus.*[63] Additionally, aminoglycosides were shown to inhibit translocation,[64,65] although the nature and relative importance of this effect remain to be explained.[66,67] Because aminoglycosides have been shown to bind in vitro to various RNA molecules, it can be expected that they interfere with various RNA-dependent regulation pathways in vivo.[52,68-70] Consequently, it is not surprising that

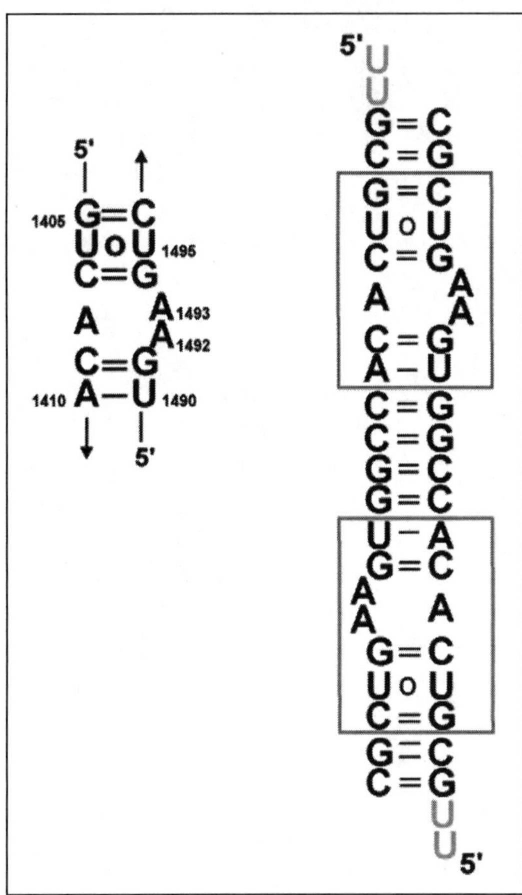

Figure 2. The secondary structure of the bacterial 16S rRNA A site with the *E. coli* numbering (left). The RNA used for crystallization is on the right.

many studies revealed that the detailed mechanism of action of aminoglycosides is a complex and delicate balance between several processes.[51] However, binding of aminoglycosides to the A site constitutes the key in their mode of action, because rRNA point mutations that preclude aminoglycoside/A-site interactions confer strong resistance to aminoglycosides.[17,71,72]

Mode of Binding of Aminoglycosides to the A Site: Stabilization of One Conformational State

Crystallographic structures of various 30S particle and minimal A-site complexes (Fig. 2) helped to visualize the mechanism of action of aminoglycosides at the ribosomal level in atomic detail. During decoding, the A site changes its conformation from an "off" conformation (with A1492 and A1493 folded into the shallow groove of the A site) to an "on" conformation (with A1492 and A1493 fully bulged out from the A site), as shown in Figure 3.[56,73-75] The electron density was inconsistent with a single conformation for both adenines A1492 and A1493 in the absence of aminoglycosides, implicating a dynamic equilibrium of that region.[73] Clear density was obtained after soaking an anticodon hairpin together with a single-stranded RNA

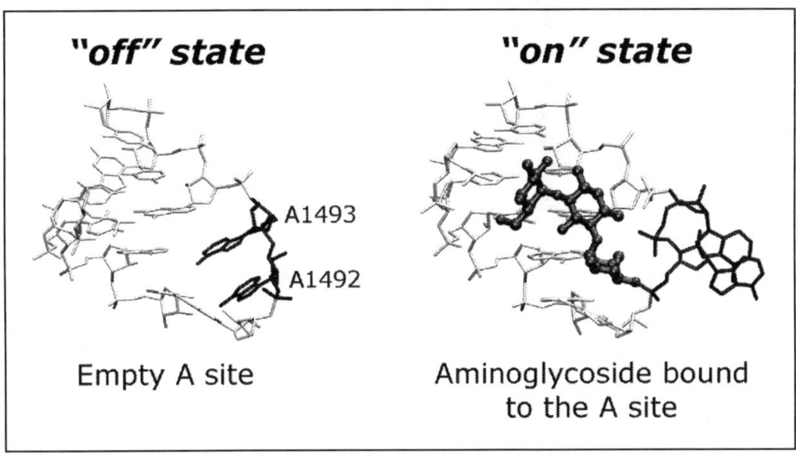

Figure 3. Views of two crystallographic states between which the A site is in dynamic equilibrium. Aminoglycosides block the A site in its "on" conformation, which is the state normally induced and stabilized by binding of cognate tRNA to a codon. The "off" state can be observed in multiple conformations (Vicens et al, in preparation). Coordinates of the two main states of the A site are extracted from PDB ID 1J5E (left) and PDB ID 1LC4 (right).

into crystals of the 30S particle.[56,74] The observed conformational change is necessary to allow A1492 and A1493 to interact specifically with the first two of the three base pairs formed by the cognate codon:anticodon interaction.[74] This structural change also provokes the transition of the ribosome from an open to a closed form that is stabilized by contacts involving the cognate tRNA and the ribosome.[56,57] Aminoglycosides lock the A site in the open conformation (Fig. 3)[8] and, by doing so, they also pay for a part of the energetic cost associated with the tRNA-dependent ribosome closure.[56,57] As a consequence, the ribosome loses its ability to discriminate cognate versus noncognate tRNA-mRNA associations.[56,58,74]

The crystal structures of an RNA double helix containing two A sites in complex with paromomycin, tobramycin, and geneticin characterized the binding mode of aminoglycosides at high resolution.[13-15] The puckered sugar ring I is inserted into the A-site helix by stacking against a guanine residue and by forming a pseudo pair with two H-bonds to the Watson-Crick sites of the universally conserved adenine 1408. As was observed in the 30S particle, this particular interaction helps to maintain adenines 1492 and 1493 in the bulged-out conformation that induces misreading.[74] The conserved 2-deoxystreptamine ring (ring II) forms similar H-bonds in the three complexes, and its binding is made possible by the adaptability created by the universally conserved U1406•U1495 pair.[72] The additional rings contact different nucleotides of the A site, depending on the substitution type of ring II. One-third of the total RNA-aminoglycoside contacts were shown to be mediated by water molecules.[13,16] Thus, overall tight packing of atoms in direct van der Waals contact is central and a prerequisite to specific recognition. Water molecules participate in the assembly by linking hydrophilic groups that belong to both components. The hydration shells around nucleic acid base pairs tend to be conserved and maintained regardless of the environment.

In recent crystal structures,[76] variable occupations of each of the two binding sites are observed (François et al., in preparation). Depending on the nature of the antibiotic, either zero, one, or two antibiotic molecules are bound per site. In these structures, when the A site is empty, only a single adenine (A1492) is bulged out whereas the other (A1493) pairs with A1408. A recent 1.7-Å crystal structure of the empty A site inserted into a different oligonucleotide shows two-state conformational disorder for A1492. In one conformation, A1492 forms a *cis* Watson-Crick pair with A1408 (with A1493 bulged out), and in the other conformation, both A1492 and A1493 bulge out and stack together.[75] The strongest evidence for a dynamic equilibrium influenced by aminoglycosides was obtained by correlating the latter crystal structure[75] to fluorescence-induced changes of A-site fragments with the bulging adenines mutated to the fluorescent 2-aminopurine.[77] The congruence between the fluorescence-induced changes and the crystal structures clearly indicated that paromomycin displaces A1492 from the internal loop, after which A1492 stacks with A1493.[75]

The Decoding Process by the Two Bulging Adenines of the A Site

Adenines 1492 and 1493 bulge out of the A-site helix to form A-minor contacts with the first two base pairs of the codon-anticodon.[8] The structures of the minimal A-site RNA complexed to various aminoglycosides display crystal packings with intermolecular contacts between the bulging adenines (equivalent to A1492 and A1493) and the shallow/minor groove of a neighboring helix, mimicking contacts that are seen in the 30S crystals.[13-15] In A-minor motifs, two adenines interact with the 3'-end of two adjacent helical Watson-Crick base pairs, such that the first A (A1492 in the A site) interacts in an antiparallel fashion with one base of the Watson-Crick pair, and the second (A1493) interacts in a parallel fashion with one base of the next Watson-Crick pair. Two important points are as follows: (1) A1493 interacts with both strands of the receptor helix, whereas A1492 interacts only with one strand; (2) A1493 forms more hydrogen bonds than A1492. For the decoding process, the general scheme is shown in Figure 4. The anticodon nucleotides, corresponding to positions 35 and 36 in the nomenclature of tRNA[Phe], are denoted ac35 and ac36. The first and second positions of the codon are denoted c1 and c2. The type I and type II names correspond to the nomenclature of reference 78 and the *trans* sugar-edge/sugar-edge or *cis* sugar-edge/sugar-edge to that of reference 79. The hydrogen bonding interactions are shown in Figure 5.

Interactions between adenines and shallow/minor grooves of RNA were first proposed during our modeling of group I introns.[18] On the basis of sequence analysis and during the three-dimensional modeling, it was noticed that the over-represented GNRA tetraloops were close to regular RNA helices, such that the last two residues (-RA-) could interact with two helically stacked base pairs R-Y and G=C. The covariation was such that when R is an A, R-Y would be G-C, and when R is a G, R-Y would be A-U. From the modeling, it was clear that the interactions can only occur in the minor/shallow groove. This type of interaction was later demonstrated using chemical probing[80,81] and mutagenesis[82] and soon afterwards by X-ray crystallography of the hammerhead ribozyme.[83,84] Two years later, the crystal structure of the

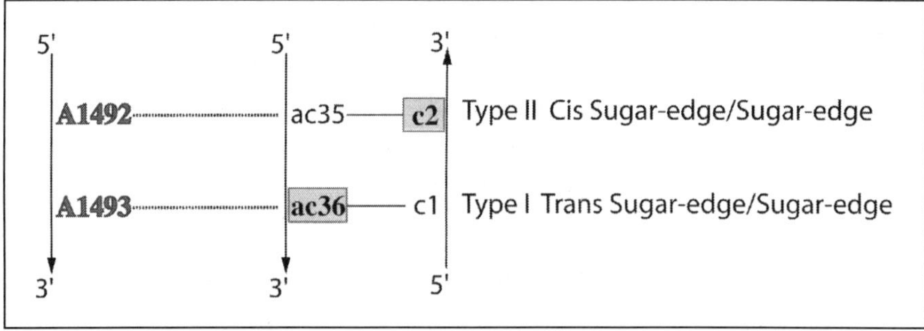

Figure 4. The schematic of the pairing scheme between A1492 and A1493 and the first two base pairs of the codon-anticodon minihelix.

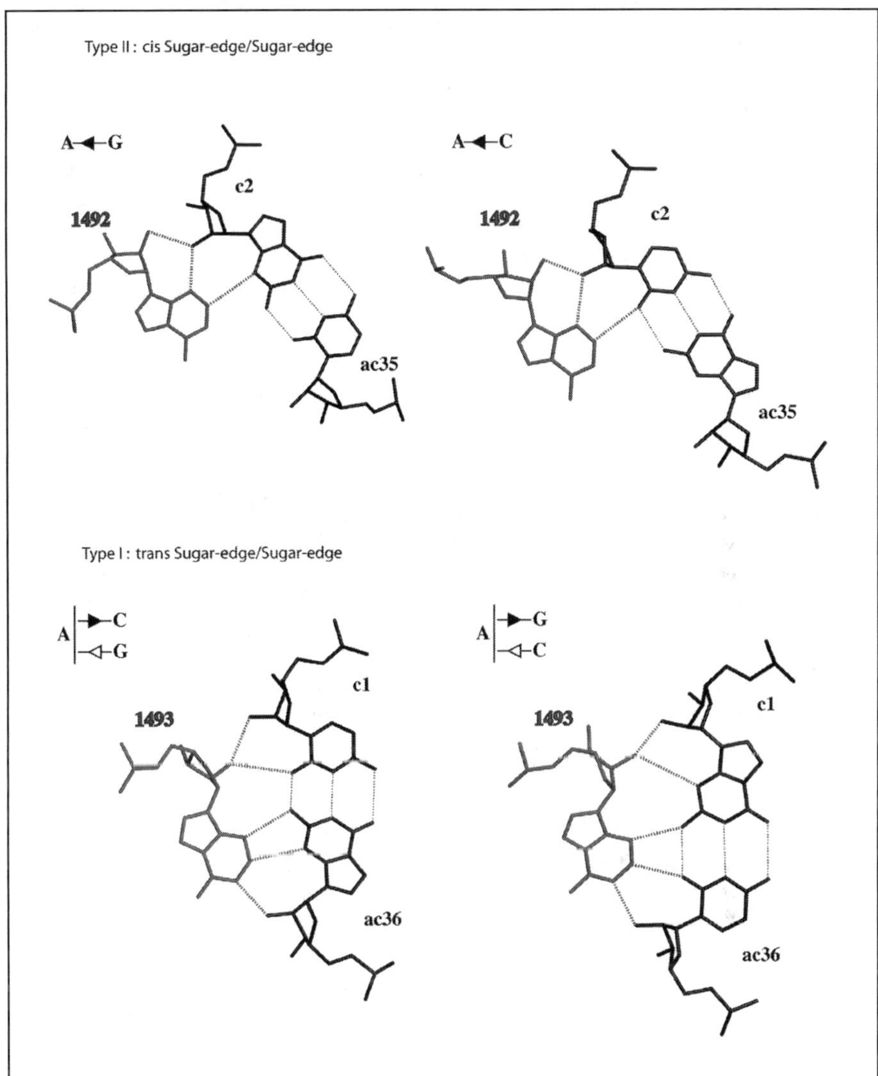

Figure 5. Atomic details of the hydrogen-bonding schemes between the bulging adenines A1492 and A1493 of the A site and the codon-anticodon helix formed between the messenger RNA and the tRNA. Notice the larger number of H-bonds formed for A1493 compared with A1492.

P4-P6 domain of the *Tetrahymena* group I intron revealed two more examples of interactions between adenines and the shallow/minor groove of RNA helices.[85,86] It revealed also a new type of RNA-RNA motif which had been previously identified by in vitro selection.[87] The universality and importance of the A-minor motifs for RNA structure were definitively established by the crystal structures of the ribosomal particles.[8,78]

Significantly for the decoding process, whereas GNRA tetraloops are able to distinguish between two consecutive helical base pairs (either G=C or A-U), two consecutive adenines in A-minor motifs bind to any combination of two stacked Watson-Crick pairs. The following two important points have been shown: (1) consecutive adenines recognize any two stacked Watson-Crick pairs without any strong bias;[88] (2) consecutive adenines have a strong preference for complementary Watson-Crick pairs as compared with noncomplementary pairs.[89] Ramakrishnan's group, using cocrystals of 30S particles in presence of codons with near-cognate tRNA anticodon stem loops (i.e., with G∘U either at the first or at the second position of the minihelix formed between codon and anticodon triplets) demonstrated those points in the case of the decoding process.[56] Crystal structures revealed that the first base of the codon interacts only via its 2'-OH group with A1493 (because the U of the G∘U wobble moves into the major/ deep groove). The situation for the second position of the codon was understandably much less defined, because most contacts between the codon nucleotide and A1492 are lost in case of a G∘U base pair, for which the U is moved away into the deep groove.

In conclusion, the most common and frequent RNA-RNA self-assembly motif is at the very core of the recognition of the codon-anticodon helix and is central to the fidelity of the ribosomal decoding process. Clearly, ribosomal RNAs do not constitute a passive platform for other active players; on the contrary, they take active part in the recognition processes that underlie translational fidelity. While the RNA parts of the 50S particle catalyze peptide bond formation,[90] the RNA parts of the 30S particle are directly and actively at the center of the recognition processes that guarantee correct codon-anticodon interactions, thereby constituting the primary factor contributing to translation fidelity. In short, the ribosome is a ribozyme[90] and, like all other known ribozymes, the ribosome uses RNA-based recognition motifs not only for catalysis and substrate release but also for decoding. These two conclusions support the RNA World hypothesis with the primordial appearance of RNA in the origin of life.[91-93] One can speculate that the observed frequent occurrence of the A-minor motif indicates an ancient, if not primordial, prevalence of those motifs in the RNA World. Therefore, it is not too surprising that this motif should be used for recognition of the codon-anticodon helix. And, following Crick's hypothesis of the frozen origin of the genetic code,[94] once such a recognition system was incorporated into the decoding process, it could not be replaced but could only be more finely tuned in the overall fidelity process.

The Importance of Targeting a Molecular Switch

The A site exists as a dynamic equilibrium between an open form (with both bases A1492 and A1493 bulging out) and a closed form (with either or both bases within the internal loop). Upon binding of a cognate tRNA to the messenger codon, the equilibrium is displaced toward the open form, with the bulging adenines forming A-minor contacts with the first two base pairs of the codon-anticodon helix. With near-cognate tRNAs, the loss in hydrogen bonding prevents the stabilization of the productive state. The binding of aminoglycosides to the A site leads to a conformation that mimicks the conformation adopted by the A site in presence of cognate tRNA-mRNA codon association, which locks it into the open conformation. Thus, the binding of aminoglycosides shunts a natural molecular switch, resulting in a loss of translation fidelity.

Similarly, recent results indicate that sparsomycin catalyzes ribosomal translocation by binding to the peptidyl-tRNA and the 23S rRNA,[10,95] a process normally performed by elongation factor G and guanosine triphosphate.[67] The effects of sparsomycin in the translocation process can be understood if sparsomycin binds preferentially to the translocated state, thereby shunting another type of control point in the translation process. Interestingly, the aminoglycoside hygromycin B, which binds mainly at the level of the conserved U1406∙U1495 pair and thus between the A and P sites, blocks ribosomal translocation without inducing misreading.[72] Because aminoglycosides such as paromomycin and tobramycin that bind to the A site also contact the U1406∙U1495 pair, one would expect that they also

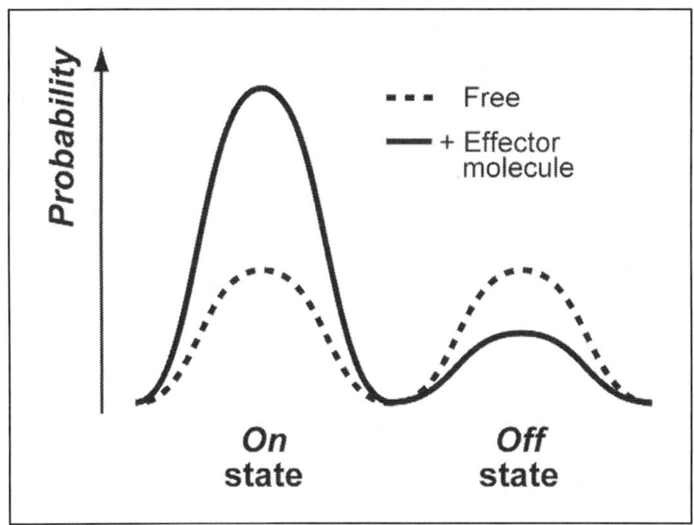

Figure 6. General scheme for the inhibition of RNA molecular switches by small molecules. The effector is shown here to stabilize the "on" state. The relative probability of occurrence of any state is proportional to exp(–E/kT)/Q, where E is the energy of that state and Q the partition function.

hinder translocation, as shown by Fredrick and Noller.[67] The study of aminoglycoside variants that do not contact the U1406•U1495 pair would be helpful in the further understanding of these connected processes.

In short, although a strong specificity might be difficult to achieve with oppositely charged molecules, targeting motifs that undergo dynamic exchange between alternative conformations should improve the biological activity of antibacterial compounds. In analogy to the aminoglycoside and sparsomycin cases, RNA targets ought to be chosen in regions that switch between alternative states, where each state—referred to as an "on" or "off" state—leads to a different biological action. One of the two states would then predominate in the presence of the small molecule (Fig. 6). Such a situation was found for the inhibition of the yeast tRNA^{Asp} aminoacylation reaction by aminoglycosides: the binding of tobramycin alters the native conformation of the tRNA so that binding of the synthetase becomes less favorable.[96]

This approach is likely to gain favor now that it is becoming apparent that several important biological mechanisms are similarly regulated by molecular switches that involve distinct RNA conformational states.[50,97,98] For example, small effectors promote up- or down-regulation of a particular protein's expression by binding to the 5'-UTR regions of the corresponding mRNA, thereby favoring a particular "on" or "off" conformation (riboswitches; see Chapter 6).[97,99,100] Recent bioinformatic algorithms enable the identification of potential conformational switches in mRNA and other regulatory RNA molecules.[101,102] Ultimately, identifying and understanding such control mechanisms in the bacterial ribosome as well as in other RNA molecules should help to discover new viral and bacterial targets for therapeutic intervention.

Perspective

The standard approaches of designing drugs to target proteins have benefited from crystal structures of complexes between inhibitors and proteins for the last 20 years. In contrast, the first crystal structures of small molecules bound to RNA pockets were solved only a few years ago. The crystal structures of the ribosomal subunits from various organisms bound to antibiotics that have been used for decades gave a strong impetus to the antibiotic research field by revealing that most of the drugs bind to defined RNA motifs. It is now becoming apparent that antibiotics target translation fidelity by altering the kinetically controlled conformational changes that are at the heart of the ribosomal machinery. In several aspects, the 16S rRNA A site is a very special target: it is at the base of the almost terminal long helix 44, which is located at the interface between the two subunits in a region devoid of proteins. The A site is the key RNA element at the core of the RNA-controlled decoding process.[57] It presents its major/deep groove to the solvent face[103] and maintains its structural properties (such as its ability to interact with antibiotics) outside the context of the ribosome.[104-106] Most of the other binding surfaces for antibiotics are scattered among residues brought close by the tertiary folding but far apart in secondary structure. The hope to analyze them in isolation is therefore tenuous. Are there other unsuspected switchable RNA targets present in the ribosome? How could we search for them?

Normal mode analysis[107,108] has been used to locate regions important for open/closed transitions in proteins in order to develop new compounds that inhibit such molecular switches.[109] This technique, when combined with further understanding of the translation steps, should help identify new RNA switches as promising targets. However, the very fact that RNA self-assembly and recognition motifs are modular and recurrent constitutes a major drawback for drug designers, due to the difficulties that must be faced to achieve specificity and avoid resistance. The developments of future therapies will undoubtedly require the concerted efforts of microbiologists, biochemists, structural biologists, bio-informaticians and organic chemists.

Acknowledgements

We thank Erik Böttger and Stephen Hanessian for numerous discussions. We are grateful to Aurélie Lescoute-Philipps for the help with Figures 4 and 5. E. Westhof acknowledges the Institut universitaire de France for support.

References

1. Gale EF, Cundliffe E, Reynolds PE et al. The molecular basis of antibiotic action. London: John Wiley & Sons Inc. 1981.
2. Walsh C. Molecular mechanisms that confer antibacterial drug resistance. Nature 2000; 406:775-81.
3. Hermann T. Chemical and functional diversity of small molecule ligands for RNA. Biopolymers 2003; 70:4-18.
4. Moazed D, Noller HF. Interaction of antibiotics with functional sites in 16S ribosomal RNA. Nature 1987; 327:389-94.
5. De Stasio E, Moazed D, Noller HF et al. Mutations in 16S ribosomal RNA disrupt antibiotic-RNA interactions. EMBO J 1989; 8:1213-6.
6. Pape T, Wintermeyer W, Rodnina MV. Conformational switch in the decoding region of 16S rRNA during aminoacyl-tRNA selection on the ribosome. Nat Struct Biol 2000; 7:104-7.
7. Brodersen DE et al. The structural basis for the action of the antibiotics tetracycline, pactamycin, and hygromycin B on the 30S ribosomal subunit. Cell 2000; 103:1143-1154.
8. Carter AP et al. Functional insights from the structure of the 30S ribosomal subunit and its interactions with antibiotics. Nature 2000; 407:340-8.
9. Auerbach T et al. Antibiotics targeting ribosomes: Crystallographic studies. Curr Drug Targets Infect Disord 2002; 2:169-86.
10. Hansen JL et al. The structures of four macrolide antibiotics bound to the large ribosomal subunit. Mol Cell 2002; 10:117-28.
11. Pioletti M et al. Crystal structures of complexes of the small ribosomal subunit with tetracycline, edeine and IF3. EMBO J 2001; 20:1829-39.

12. Schlünzen F et al. Structural basis for the interaction of antibiotics with the peptidyl transferase centre in eubacteria. Nature 2001; 413:814-21.

13. Vicens Q, Westhof E. Crystal structure of paromomycin docked into the eubacterial ribosomal decoding A site. Structure 2001; 9:647-658.

14. Vicens Q, Westhof E. Crystal structure of a complex between the aminoglycoside tobramycin and an oligonucleotide containing the ribosomal decoding A-site. Chem Biol 2002; 9:747-755.

15. Vicens Q, Westhof E. Crystal structure of geneticin bound to a bacterial 16S ribosomal RNA A site oligonucleotide. J Mol Biol 2003; 326:1175-1188.

16. Vicens Q, Westhof E. Molecular recognition of aminoglycoside antibiotics by ribosomal RNA and resistance enzymes: An analysis of X-ray crystal structures. Biopolymers 2003; 70:42-57.

17. Vicens Q, Westhof E. RNA as a drug target: The case of aminoglycosides. ChemBioChem 2003; 4:1018-23.

18. Michel F, Westhof E. Modelling of the three-dimensional architecture of group I catalytic introns based on comparative sequence analysis. J Mol Biol 1990; 216:585-610.

19. Massire C, Jaeger L, Westhof E. Derivation of the three-dimensional architecture of bacterial ribonuclease P RNAs from comparative sequence analysis. J Mol Biol 1998; 279:773-93.

20. Tinoco Jr I, Bustamante C. How RNA folds. J Mol Biol 1999; 293:271-81.

21. Batey RT, Rambo RP, Doudna JA. Tertiary motifs in RNA structure and folding. Angew Chem Int Ed 1999; 38:2326-2343.

22. Rangan P, Masquida B, Westhof E et al. Assembly of core helices and rapid tertiary folding of a small bacterial group I ribozyme. Proc Natl Acad Sci USA 2003; 100:1574-9.

23. Brion P, Westhof E. Hierarchy and dynamics of RNA folding. Annu Rev Biophys Biomol Struct 1997; 26:113-37.

24. Hammann C, Lilley DM. Folding and activity of the hammerhead ribozyme. ChemBioChem 2002; 3:690-700.

25. Khvorova A, Lescoute A, Westhof E et al. Sequence elements outside the hammerhead ribozyme catalytic core enable intracellular activity. Nat Struct Biol 2003; 10:708-12.

26. Wilson TJ, Lilley DM. Metal ion binding and the folding of the hairpin ribozyme. RNA 2002; 8:587-600.

27. Doudna JA, Cech TR. The chemical repertoire of natural ribozymes. Nature 2002; 418:222-8.

28. FerreD'Amare AR, Doudna JA. Crystallization and structure determination of a hepatitis delta virus ribozyme: Use of the RNA-binding protein U1A as a crystallization module. J Mol Biol 2000; 295:541-56.

29. Leontis NB, Westhof E. Analysis of RNA motifs. Curr Opin Struct Biol 2003; 13:300-8.

30. Westhof E, Auffinger P. In: Meyers RA, ed. Encyclopedia of Analytical Chemistry. Chichester: John Wiley & Sons Ltd, 2000:5222-5232.

31. Hermann T, Westhof E. Rational drug design and high-throughput techniques for RNA targets. Comb Chem High Throughput Screen 2000; 3:219-34.

32. Leontis NB, Westhof E. A common motif organizes the structure of multi-helix loops in 16 S and 23 S ribosomal RNAs. J Mol Biol 1998; 283:571-83.

33. Westhof E, Masquida B, Jaeger L. RNA tectonics: Towards RNA design. Fold & Des 1996; 1:R78-R88.

34. Karplus M, McCammon JA. The internal dynamics of globular proteins. CRC Crit Rev Biochem 1981; 9:293-349.

35. Westhof E et al. Correlation between segmental mobility and the location of antigenic determinants in proteins. Nature 1984; 311:123-6.

36. Tainer JA et al. The reactivity of anti-peptide antibodies is a function of the atomic mobility of sites in a protein. Nature 1984; 312:127-34.

37. Westhof E, Roder O, Croneiss I et al. Ribose conformations in the common purine(beta)ribosides, in some antibiotic nucleosides, and in some isopropylidene derivatives: A comparison. Z Naturforsch [C] 1975; 30:131-40.

38. Woodson SA, Crothers DM. Preferential location of bulged guanosine internal to a G.C tract by ^{1}H NMR. Biochemistry 1988; 27:436-45.

39. Woodson SA, Crothers DM. Proton nuclear magnetic resonance studies on bulge-containing DNA oligonucleotides from a mutational hot-spot sequence. Biochemistry 1987; 26:904-12.

40. Ehresmann C et al. A pseudoknot is required for efficient translational initiation and regulation of the Escherichia coli rpsO gene coding for ribosomal protein S15. Biochem Cell Biol 1995; 73:1131-40.

41. Benard L, Philippe C, Ehresmann B et al. Pseudoknot and translational control in the expression of the S15 ribosomal protein. Biochimie 1996; 78:568-76.

42. Pan J, Woodson SA. Folding intermediates of a self-splicing RNA: Mispairing of the catalytic core. J Mol Biol 1998; 280:597-609.

43. Pan J, Woodson SA. The effect of long-range loop-loop interactions on folding of the Tetrahymena self-splicing RNA. J Mol Biol 1999; 294:955-65.

44. Emerick VL, Woodson SA. Self-splicing of the Tetrahymena prerRNA is decreased by misfolding during transcription. Biochemistry 1993; 32:14062-7.

45. Lilley DM. Structures of helical junctions in nucleic acids. Q Rev Biophys 2000; 33:109-59.

46. Hohng S et al. Conformational flexibility of four-way junctions in RNA. J Mol Biol 2004; 336:69-79.

47. Goody TA, Lilley DM, Norman DG. The chirality of a four-way helical junction in RNA. J Am Chem Soc 2004; 126:4126-7.

48. Goody TA, Melcher SE, Norman DG et al. The kink-turn motif in RNA is dimorphic, and metal ion-dependent. RNA 2004; 10:254-64.

49. Klein DJ, Schmeing TM, Moore PB et al. The kink-turn: A new RNA secondary structure motif. EMBO J 2001; 20:4214-21.

50. Bashan A et al. Structural basis of the ribosomal machinery for Peptide bond formation, translocation, and nascent chain progression. Mol Cell 2003; 11:91-102.

51. Wright GD, Berghuis AM, Mobashery S. In: Rosen BP, Mobashery S, eds. Resolving the Antibiotic Paradox: Progress in Understanding Drug Resistance and Development of New Antibiotics. New York: Plenum Press, 1998:27-69.

52. Walter F, Vicens Q, Westhof E. Aminoglycoside-RNA interactions. Curr Opin Chem Biol 1999; 3:694-704.

53. Ritter TK, Wong CH. Carbohydrate-based antibiotics: A new approach to tackling the problem of resistance. Angew Chem Int Ed 2001; 40:3508-3533.

54. Davies J, Gorini L, Davis BD. Misreading of RNA codewords induced by aminoglycoside antibiotics. Mol Pharmacol 1965; 1:93-106.

55. Pape T, Wintermeyer W, Rodnina M. Induced fit in initial selection and proofreading of aminoacyl-tRNA on the ribosome. EMBO J 1999; 18:3800-7.

56. Ogle JM, Murphy FV, Tarry MJ et al. Selection of tRNA by the ribosome requires a transition from an open to a closed form. Cell 2002; 111:721-32.

57. Ogle JM, Carter AP, Ramakrishnan V. Insights into the decoding mechanism from recent ribosome structures. Trends Biochem Sci 2003; 28:259-66.

58. Rodnina MV, Wintermeyer W. Fidelity of aminoacyl-tRNA selection on the ribosome: Kinetic and structural mechanisms. Annu Rev Biochem 2001; 70:415-35.

59. Gromadski KB, Rodnina MV. Streptomycin interferes with conformational coupling between codon recognition and GTPase activation on the ribosome. Nat Struct Mol Biol 2004; 11:316-22.

60. Hausner TP, Geigenmuller U, Nierhaus KH. The allosteric three-site model for the ribosomal elongation cycle. New insights into the inhibition mechanisms of aminoglycosides, thiostrepton, and viomycin. J Biol Chem 1988; 263:13103-11.

61. Dinos G, et al. Dissecting the ribosomal inhibition mechanisms of edeine and pactamycin: The universally conserved residues G693 and C795 regulate P-site RNA binding. Mol Cell 2004; 13:113-24.

62. Mehta R, Champney WS. 30S ribosomal subunit assembly is a target for inhibition by aminoglycosides in Escherichia coli. Antimicrob Agents Chemother 2002; 46:1546-9.

63. Mehta R, Champney WS. Neomycin and paromomycin inhibit 30S ribosomal subunit assembly in Staphylococcus aureus. Curr Microbiol 2003; 47:237-43.

64. Cabanas MJ, Vazquez D, Modolell J. Inhibition of ribosomal translocation by aminoglycoside antibiotics. Biochem Biophys Res Commun 1978; 83:991-7.

65. Davies J, Davis BD. Misreading of ribonucleic acid code words induced by aminoglycoside antibiotics. J Biol Chem 1968; 243:3312-3316.
66. Phelps SS, Jerinic O, Joseph S. Universally conserved interactions between the ribosome and the anticodon stem-loop of A site tRNA important for translocation. Mol Cell 2002; 10:799-807.
67. Fredrick K, Noller HF. Catalysis of ribosomal translocation by sparsomycin. Science 2003; 300:1159-62.
68. Schroeder R, Waldsich C, Wank H. Modulation of RNA function by aminoglycoside antibiotics. EMBO J 2000; 19:1-9.
69. Tor Y. RNA and the small molecule world. Angew Chem Int Ed 1999; 38:1579-1582.
70. Michael K, Tor Y. Designing novel RNA binders. Chem Eur J 1998; 4:2091-2098.
71. Böttger EC, Springer B, Prammananan T et al. Structural basis for selectivity and toxicity of ribosomal antibiotics. EMBO Rep 2001; 2:318-23.
72. Pfister P, Hobbie S, Vicens Q et al. The molecular basis for A-site mutations conferring aminoglycoside resistance: Relationship between ribosomal susceptibility and X-ray crystal structures. ChemBioChem 2003; 4:1078-88.
73. Wimberly BT et al. Structure of the 30S ribosomal subunit. Nature 2000; 407:327-39.
74. Ogle JM et al. Recognition of cognate transfer RNA by the 30S ribosomal subunit. Science 2001; 292:897-902.
75. Shandrick S et al. Monitoring molecular recognition of the ribosomal decoding site. Angew Chem Int Ed 2004; 43:3177-82.
76. Russell RJ, Murray JB, Lentzen G et al. The complex of a designer antibiotic with a model aminoacyl site of the 30S ribosomal subunit revealed by X-ray crystallography. J Am Chem Soc 2003; 125:3410-1.
77. Kaul M, Barbieri CM, Pilch DS. Fluorescence-based approach for detecting and characterizing antibiotic-induced conformational changes in ribosomal RNA: Comparing aminoglycoside binding to prokaryotic and eukaryotic ribosomal RNA sequences. J Am Chem Soc 2004; 126:3447-53.
78. Nissen P, Ippolito JA, Ban N et al. RNA tertiary interactions in the large ribosomal subunit: The A-minor motif. Proc Natl Acad Sci USA 2001; 98:4899-903.
79. Leontis NB, Westhof E. Geometric nomenclature and classification of RNA base pairs. RNA 2001, 7:499-512.
80. Murphy FL, Cech TR. An independently folding domain of RNA tertiary structure within the Tetrahymena ribozyme. Biochemistry 1993; 32:5291-300.
81. Murphy FL, Cech TR. GAAA tetraloop and conserved bulge stabilize tertiary structure of a group I intron domain. J Mol Biol 1994; 236:49-63.
82. Jaeger L, Michel F, Westhof E. Involvement of a GNRA tetraloop in long-range RNA tertiary interactions. J Mol Biol 1994; 236:1271-6.
83. Pley HW, Flaherty KM, McKay DB. Three-dimensional structure of a hammerhead ribozyme. Nature 1994; 372:68-74.
84. Pley HW, Flaherty KM, McKay DB. Model for an RNA tertiary interaction from the structure of an intermolecular complex between a GAAA tetraloop and an RNA helix. Nature 1994; 372:111-113.
85. Cate JH et al. Crystal structure of a group I ribozyme domain: Principles of RNA packing. Science 1996; 273:1678-85.
86. Cate JH et al. RNA tertiary structure mediation by adenosine platforms. Science 1996; 273:1696-9.
87. Costa M, Michel F. Frequent use of the same tertiary motif by self-folding RNAs. EMBO J 1995; 14:1276-85.
88. Doherty EA, Batey RT, Masquida B et al. A universal mode of helix packing in RNA. Nat Struct Biol 2001; 8:339-43.
89. Battle DJ, Doudna JA. Specificity of RNA-RNA helix recognition. Proc Natl Acad Sci USA 2002; 99:11676-81.
90. Nissen P, Hansen J, Ban N et al. The structural basis of ribosome activity in peptide bond synthesis. Science 2000; 289:920-30.
91. Crick FH. The origin of the genetic code. J Mol Biol 1968; 38:367-79.

92. Orgel LE. Evolution of the genetic apparatus. J Mol Biol 1968; 38:381-93.

93. Gilbert W. The RNA world. Nature 1986; 319-618.

94. Crick FH. The genetic code—yesterday, today, and tomorrow. Cold Spring Harb Symp Quant Biol 1966; 31:1-9.

95. Hansen JL, Schmeing TM, Moore PB et al. Structural insights into peptide bond formation. Proc Natl Acad Sci USA 2002; 99:11670-5.

96. Walter F, Pütz J, Giegé R et al. Binding of tobramycin leads to conformational changes in yeast tRNA^{Asp} and inhibition of aminoacylation. EMBO J 2002; 21:760-8.

97. Nahvi A et al. Genetic control by a metabolite binding mRNA. Chem Biol 2002; 9:1043.

98. Mandal M, Boese J, Barrick JE et al. Riboswitches control fundamental biochemical pathways in Bacillus subtilis and other bacteria. Cell 2003; 113:577-586.

99. Winkler W, Nahvi A, Breaker RR. Thiamine derivatives bind messenger RNAs directly to regulate bacterial gene expression. Nature 2002; 419:952-6.

100. Mandal M, Breaker RR. Gene regulation by riboswitches. Nat Rev Mol Cell Biol 2004; 5:451-63.

101. Barash D. Second eigenvalue of the Laplacian matrix for predicting RNA conformational switch by mutation. Bioinformatics 2004; 20:1861-9.

102. Voss B, Meyer C, Giegerich R. Evaluating the predictability of conformational switching in RNA. Bioinformatics 2004; 20:1573-82.

103. Yusupov MM et al. Crystal structure of the ribosome at 5.5 A resolution. Science 2001; 292:883-96.

104. Purohit P, Stern S. Interactions of a small RNA with antibiotics and RNA ligands of the 30S subunit. Nature 1994; 370:659-662.

105. Recht MI, Fourmy D, Blanchard SC et al. RNA sequence determinants for aminoglycoside binding to an A-site rRNA model oligonucleotide. J Mol Biol 1996; 262:421-436.

106. Miyaguchi H, Narita H, Sakamoto K et al. An antibiotic-binding motif of an RNA fragment derived from the A-site related region of Escherichia coli 16S RNA. Nucleic Acids Res 1996; 24:3700-3706.

107. Tama F, Sanejouand YH. Conformational change of proteins arising from normal mode calculations. Protein Eng 2001; 14:1-6.

108. Tama F, Valle M, Frank J et al. Dynamic reorganization of the functionally active ribosome explored by normal mode analysis and cryo-electron microscopy. Proc Natl Acad Sci USA 2003; 100:9319-23.

109. Delarue M, Sanejouand YH. Simplified normal mode analysis of conformational transitions in DNA-dependent polymerases: The elastic network model. J Mol Biol 2002; 320:1011-24.

Index

S

Selection 5, 7-13, 25, 33, 37, 39, 40, 42, 45, 49-51, 55, 57, 58, 61, 62, 66-68, 91, 98, 110, 111, 115
Self-splicing 4, 7, 9, 19, 25, 40, 42, 76
Sequence analysis 114
snRNP 84, 85
Sparsomycin 110, 116, 117
Spectinomycin 110
StpA 76, 82, 83
Superoxide dismutase (SOD) 80, 81

T

Tandem aptamer 99-101
TAR 42
Tat 42, 57, 58
Terminator 91, 94-96, 99, 100
Tetracycline 110
Tetrahymena 7, 9, 13, 40-42, 79, 115
Tetraloop 7, 40-42, 114, 115
Theophylline 5, 6, 12, 13, 19, 20, 101
Thrombin 15, 57, 58, 60
Tobramycin 108, 110, 113, 116, 117

V

Varkud satellite (VS) 4, 25

X

X-ray 5, 38, 93, 109, 114

Z

Zinc finger 83